Compliance Guide for the Concentrated Aquatic Animal Production Point Source Category

United States Environmental Protection Agency
Office of Water
Washington, DC 20460
(4303T)

EPA-821-B-05-001
March 2006

Compliance Guide for the Concentrated Aquatic Animal Production Point Source Category

Engineering and Analysis Division
Office of Science and Technology
U.S. Environmental Protection Agency
Washington, DC 20460

March 2006

Acknowledgments

The Engineering and Analysis Division, Office of Science and Technology has prepared this compliance guidance and approved it for publication. Tetra Tech, Inc., Contract 68-C-99-263, supported the preparation of the guidance under the direction and review of the Office of Science and Technology.

Technical experts in the concentrated aquatic animal production (CAAP) industry prepared the document *Best Management Practices for Flow-Through, Net-Pen, Recirculating, and Pond Aquaculture Systems* by Tucker, C., S. Belle, C. Boyd, G. Fornshell, J. Hargreaves, S. LaPatra, S. Summerfelt, and P. Zajicek (2003). It provided valuable background information for EPA in the development of this compliance guide.

Disclaimer

The discussion in this document is intended solely as guidance. The statutory provisions and regulations of the U.S. Environmental Protection Agency (EPA) described in this document contain legally binding requirements. This document is not a regulation itself, nor does it change or substitute for those provisions and regulations. Thus, it does not impose legally binding requirements on EPA, States, or the regulated community. This guidance does not confer legal rights or impose legal obligations upon any member of the public.

While EPA has made every effort to ensure the accuracy of the discussion in this guidance, the obligations of the regulated community are determined by statutes, regulations, or other legally binding requirements. In the event of a conflict between the discussion in this document and any statute or regulation, this document would not be controlling.

The general descriptions provided here may not apply to particular situations based upon the circumstances. Interested parties are free to raise questions and objections about the substance of this guidance and the appropriateness of the application of this guidance to a particular situation. EPA and other decision-makers retain the discretion to adopt approaches on a case-by-case basis that differ from those described in this guidance where appropriate.

Mention of trade names or commercial products does not constitute an endorsement or recommendation for their use.

This document may be revised periodically without public notice. EPA welcomes public input on this document at any time.

Neither the United States government nor any of its employees, contractors, subcontractors, or other employees makes any warranty, expressed or implied, or assumes any legal liability or responsibility for any third party's use of, or the results of such use of, any information, apparatus, product, or process discussed in this report, or represents that its use by such a third party would not infringe on privately owned rights.

Table of Contents

Appendices

Chapter 1: Introduction

On June 30, 2004, the U.S. Environmental Protection Agency (EPA) completed regulations under the Clean Water Act (CWA) establishing Effluent Limitations Guidelines (ELGs) and New Source Performance Standards for the Concentrated Aquatic Animal Production (CAAP) Point Source Category. The regulations contain requirements for wastewater discharges that must be met by new and existing CAAP facilities discharging directly to U.S. waters. The CWA establishes a comprehensive program for protecting the Nation's waters. Among its core provisions, the CWA generally prohibits the discharge of pollutants from a point source to waters except as authorized by a National Pollutant Discharge Elimination (NPDES) permit. Direct dischargers must comply with effluent limits in NPDES permits. EPA's NPDES regulations define a hatchery, fish farm, or other facility as a CAAP and therefore subject to the NPDES permit program. The regulation defines a CAAP by, among other things, the size of the operation and frequency of discharge.

Those CAAP facilities subject to the ELGs must develop and maintain a best management practice (BMP) plan describing how they will achieve the ELG requirements. The CAAP must certify in writing to the permitting authority that a BMP plan has been developed and make the plan available to the permitting authority upon request. EPA did not revise the National Pollutant Discharge Elimination System (NPDES) regulation, as it applies to CAAPs.

EPA has produced this document to help CAAP facility owners/operators and NPDES permit writers to understand and comply with the NPDES and ELGs regulations. Background information about aquatic animal production facilities (e.g., system types, wastewater treatment) is available from EPA's Technical Development Document for the Final Effluent Limitations Guidelines and New Source Performance Standards for the Concentrated Aquatic Animal Production Point Source Category, available at http://epa.gov/guide/aquaculture/tdd/final.htm.

CAAP facilities subject to the ELGs are defined as facilities (flow-through, recirculating, and net pen) that produce 100,000 pounds or more of aquatic animals per year.

EPA is continually improving its rules, policies, compliance programs, and outreach efforts, so some of the information in this guide might have changed since it was published. You can find out whether EPA has updated this guide by checking EPA's website for the CAAP ELGs at http://epa.gov/guide/aquaculture.

How do I use the symbols contained in this document?

In many of the chapters, you will see symbols, which denote the following:

- Flow-through
- Recirculating
- Net Pen[1]

---

[1] Whenever the term net pen is used in this guidance, it also refers to cages that function like net pens.

Use the symbols for flow-through, recirculating, and net pens as a guide to determine which BMPs and/or paragraphs may apply to which system type. The symbol ⌂ denotes specific legal references.

The permit writer's version of the guidance document will contain the following symbols:

Permit writer begin

Permit writer end

Use these symbols to identify the beginning and ending of additional information or sections included specifically for permit writers.

Who should use this guide?

You should use this guide if you own or operate a CAAP facility or if you are a permit writer. It will help you to understand the June 2004, CAAP ELGs and how it relates to the NPDES regulations. Owners or operators of a CAAP can use this guide to determine if their operation is a facility subject to the ELGs. Permit writers may use this guidance to obtain information on the permitting requirements for CAAPs.

Facilities that are not covered by this rule (flow-through, recirculating, and net pen systems that produce less than 100,000 pounds of aquatic animals per year and other systems, such as ponds) may benefit from using this guidance to help improve facility operation (i.e., through feed management, materials storage, etc.) and reduce pollutant discharges.

 Permit writers may use this guidance to obtain information on permitting requirements for CAAPs. The

Guidance reflects information from the current NPDES Program and final ELGs regulations (signed on June 30, 2004 and published in the Federal Register on August 23, 2004).

The guidance assumes the permit writer has working knowledge of how to develop NPDES permits. Permit writers should also be familiar with applicable state voluntary wastewater control programs as well as regulatory programs, and how these programs relate to the federal or state NPDES program. Appendix I provides a variety of additional resources that permit writers may wish to use to increase their understanding of practices used at CAAP facilities. In addition, the guidance discusses the circumstances under which CAAP owners or operators should submit a Notice of Intent (NOI) to seek coverage under an existing NPDES general permit or apply for an NPDES individual permit.

While this guidance is limited to the development and issuance of NPDES permits for CAAPs, it is important for the permit writer to recognize that there are other NPDES program requirements that may be applicable to CAAPs. For example, discharges of storm water associated with construction activity at, or construction of, CAAPs that disturb one acre of land or more may be subject to NPDES storm water permit requirements. These requirements address activities associated with the construction of CAAPs, including clearing, grading, and excavation, but do not address discharges associated with the operation of the facility, which are addressed in the NPDES CAAP permit. Therefore, it is generally in the interest of the permitting authority and the CAAP operator to administer storm water permits for construction separately from NPDES CAAP permits. Another NPDES permitting requirement that may apply to a facility includes requirements for package plants used to treat septic waste at a facility. Additionally, if a facility has a large laboratory for fish health and diagnostic services, the lab part of the facility may be required to obtain a separate NPDES permit for their

wastewater (septic and laboratory wastes). These are only example of other types of NPDES permitting requirements that may apply at a facility.

Who is in charge of the CAAP permitting program where I live – EPA or the state?

EPA has approved most states to run their own regulatory and permitting programs for CAAPs. If EPA has approved your state, the state is the permitting authority and will issue a permit for your CAAP facility. EPA has not approved Alaska, Idaho, Massachusetts, New Hampshire, and New Mexico to permit CAAPs. In those states, D.C., tribal lands, and in all territories except the U.S. Virgin Islands, EPA is the permitting authority and will issue permits for CAAPs. Note that in some cases, EPA may still regulate some types of CAAPs even if your state has been delegated NPDES permitting authority. In these cases, the state permitting authority will direct you to the appropriate EPA contact.

Contact information for your permitting authority is available in Appendix A. Also refer to "Do other laws regulate CAAPs?" in Chapter 2 of this guide. It describes how your state, county, or town might have additional legal requirements that apply to you and that go beyond the requirements described in this guide.

What does this guide cover?

This guide describes EPA's regulations for CAAPs (NPDES and ELGs requirements), which govern whether your operation is a CAAP, whether you need a permit, and what the permit will require. State permitting authorities use EPA's regulations as a starting point but often add their own requirements in NPDES permits. You should always check with your permitting authority to see what the requirements are in your state and to find out exactly what you have to do. Appendix A contains information on how to contact your permitting authority.

This guide also provides information to help facilities develop a BMP plan for their facility, as required by the CAAP ELGs, and describes a number of BMPs that facilities may use

> Always check with your permitting authority to find out exact requirements for your facility. Your state may have more requirements or more specific requirements than the CAAP Regulations.

to achieve the requirements of the CAAP ELGs. The guide also provides example forms and logs that facility owners or operators may use to comply with the requirements of the CAAP ELGs, as well as the NPDES application form. Facilities not subject to the ELGs may also use the information in this guide to improve practices at their facility.

Finally, this guide provides permit writers with the specific permitting requirements for CAAPs, information about how CAAPs are defined, who must seek coverage under an NPDES permit, the elements of an NPDES permit for a CAAP, and other considerations when developing a permit.

How should I use this guide?

If you are a facility owner or operator, you can use this guide to determine how to comply with the requirements of the CAAP

ELGs. Read Chapter 2 ("What is the CAAP Regulation?") for basic information on the guidelines and NPDES permitting process. Chapter 3 ("Does the CAAP Regulation Affect Me?") provides information about how the CAAP regulation affects you. Read Chapter 4 ("What Do I Need to Know About NPDES Permits?) for guidance on applying for a permit. Chapter 5 discusses the requirements of an NPDES permit.

For specific guidance on complying with the CAAP regulations, refer to the following chapters:

- Chapter 6 – General Reporting Requirements for Flow-through, Recirculating, and Net Pen Facilities
- Chapter 7 – Narrative Requirements for Flow-through, Recirculating, and Net Pen Facilities
- Chapter 8 – Writing and Certifying a BMP Plan
- Chapter 9 – Solids Control for Flow-through and Recirculating Facilities
- Chapter 10 – Material Storage for Flow-through, Recirculating, and Net Pen Facilities
- Chapter 11 – Maintenance for Flow-through, Recirculating, and Net Pen Facilities
- Chapter 12 – Record-keeping for Flow-through, Recirculating, and Net Pen Facilities
- Chapter 13 – Perform Training for Flow-through, Recirculating, and Net Pen Facilities
- Chapter 14 – Feed Management for Net Pen Facilities
- Chapter 15 – Waste Collection and Disposal; Transport or Harvest Discharge; and Carcass Removal for Net Pens

Refer to the appendices for the following information:

- Appendix A: State Permitting Authorities/Departments of Environmental Protection
- Appendix B: Natural Resources Agencies Associated with Fisheries
- Appendix C: Frequently Asked Questions
- Appendix D: Code of Federal Regulations (40 CFR, Parts 122.24 and 451)
- Appendix E: BMP Plans
- Appendix F: BMP Certification Form
- Appendix G: State BMP Programs
- Appendix H: National Association of State Aquaculture Coordinators (NASAC), Cooperative Extension Services, and Sea Grant Information
- Appendix I: Additional Resources
- Appendix J: Glossary
- Appendix K: NPDES Permit Applications
- Appendix L: Applicability Matrix
- Appendix M: General Reporting Forms
  o Example Written Report – Participating in an INAD Study
  o Checklist for Oral Report for INAD and Extralabel Drug Use
  o Example Written Report – INAD and Extralabel Drug Use
  o Checklist for Oral Report of Failure or Damage to Structure of Containment Systems
  o Example Written Report – Failure or Damage to Structure of Containment Systems
  o Checklist for Oral Report of Spills of Drugs, Pesticides, and Feed
  o Example Written Report – Spills of Drugs, Pesticides, and Feed
- Appendix N: Feed Conversion Ratios Log
- Appendix O: Spills and Leaks Log

- Appendix P: Inspection and Maintenance: Logs
- Appendix Q: Cleaning Log
- Appendix R: Record-keeping Checklist
- Appendix S: Employee Training Log
- Appendix T: Carcass Removal Log
- Appendix U: FDA Labeling
- Appendix V: SDAFS BMP Plan

If you have trouble understanding any of the information in this guide, ask your permitting authority for help. You may also contact EPA's Offices of Science and Technology (Engineering and Analysis Division) and Wastewater Management (NPDES Permitting Program Branch).

EPA Contacts

Office of Science and Technology (OST), Engineering and Analysis Division (EAD)
202-566-1000

Office of Wastewater Management (OWM)
NPDES Permitting Program Branch
202-564-9545

The guide also provides permit writers with references for additional information. Permit writers may refer to the appendices listed in the previous section of this guide for additional information. Other good sources of information may be available from state agencies (e.g., Departments of Natural Resources, Agriculture, or Environmental Protection), Cooperative Extension Services, and Sea Grant.

How can I get a copy of the federal regulations?

You can obtain a copy of the federal CAAP regulation from any of the following:

- Appendix D of this document.
- View or download the text of the federal regulation as it appears in the Federal Register on EPA's ELGs website for the CAAP rule at http://epa.gov/guide/aquaculture/.
- Order the federal regulation and supporting documents from EPA's National Service Center of Environmental Publications (NSCEP).

National Service Center for Environmental Publications

Phone: 1-800-490-9198
Fax: 513-489-8695
e-mail: ncepimal@one.ent
Web: http://www.epa.gov/ncepihom
Mail: U.S. EPA/NSCEP
           P.O. Box 42419
           Cincinnati, Ohio 45242-0419

Your state might have other regulations that apply to you. Contact your permitting authority to find out how to get a copy of your state's CAAP regulations.

Supporting documents for the CAAP regulation include the Technical Development Document and the Economic and Environmental Benefits Document. They are available from NSCEP or at http://epa.gov/guide/aquaculture.

In this guide, EPA has tried to explain the regulatory language in clear, simple terms. Some of the guide's explanations are general and might not contain all the details from the regulation. Contact your permitting authority for more information on the specific regulations that apply to you. You can find contact information for your permitting authority in Appendix A of this guide.

Chapter 2: What is the CAAP Regulation?

This guide covers the requirements in the June 2004 rule for CAAP facilities. The regulation does not revise the current National NPDES Permit Regulation for CAAPs (40 CFR 122.24 and Appendix C). It does however, establish the ELGs for CAAPs (40 CFR 451).

What is the NPDES Program?

The NPDES Program was created under the federal Clean Water Act to protect and improve water quality by regulating point source dischargers. Point source dischargers are operations that *discharge pollutants* from *discrete conveyances* directly into *waters of the United States*. Point source dischargers are generally regulated by NPDES permits (40 CFR §122.2).

> *A **discharge**, in general, is the flow of treated or untreated wastewater from a facility to waters of the United States.*

> *The CWA defines **pollutant** as dredged spoil, solid waste, incinerator residue, sewage, garbage, sewage sludge, munitions, chemical wastes, biological materials, radioactive materials, heat, wrecked or discarded equipment, rock, sand, cellar dirt and industrial, municipal, and agricultural waste discharged into water.*

An NPDES permit:

- Identifies outfall points from which a facility discharges wastewater to surface waters.
- Sets requirements to protect the quality of surface water (such as pollutant concentration limits, management practices, and record-keeping requirements) that the discharger must meet.
- Allows an operation to discharge pollutants as long as the operation meets the requirements in the permit.

> *A **discrete conveyance**, in general, is any single, identifiable way for pollutants to be carried or transferred to waters, such as a pipe, ditch, or channel.*

Generally, if a facility discharges pollutants without having a permit, or has a permit but does not meet the requirements, it is violating the Clean Water Act. The owner or operator of the facility could be subject to enforcement actions such as fines.

Under the Clean Water Act, CAAPs are defined as point source dischargers. Refer to Chapter 3 of this document ("Does the CAAP Regulation Affect Me?") for a description of how EPA has defined CAAPs.

> *Every facility that meets EPA's definition of a CAAP has a duty to apply for a permit. EPA recommends that CAAP owners or operators that do not discharge should contact their permitting authority for assistance. For more information, refer to "How do I know I am covered by these regulations" in Chapter 3 of this guide.*

What are the Effluent Limitations Guidelines for CAAPs?

ELGs are national standards for wastewater discharges to surface waters and publicly owned treatment works (municipal sewage treatment plants). EPA develops ELGs for categories of existing sources and new sources under the Clean Water Act. The standards are technology-based (i.e. they are based on the performance of treatment, control technologies, and practices).

EPA completed ELGs for the CAAP industry on June 30, 2004. These ELGs are used by permitting authorities to set permit requirements for individual facilities. The requirements of the CAAP ELGs are included directly into an individual permit. In the case of CAAPs, the ELGs require *management practices and record-keeping activities*, rather than numerical limits called "discharge limits." Your state permitting authority may also set additional requirements that are needed to protect water quality or other requirements that apply under state or local law. Appendix A contains a list of permitting authorities. A summary of the regulation is available in Tables 1 through 3, at the end of this chapter.

> Note that management practices are general requirements (e.g., solids control) and facilities may choose how to achieve them. For example, solids control can be achieved through feed management and/or proper operation and maintenance of solids treatment systems. The rule does not require any specific measures to achieve solids control.

Why is this regulation important?

This regulation is important in reducing discharges of conventional pollutants (mainly total suspended solids), non-conventional pollutants (e.g., nutrients, drugs, and chemicals), and to a lesser extent, toxic pollutants (metals and PCBs) from CAAP facilities covered by the regulation.

EPA estimated that implementation of the ELGs will result in reducing the discharge of total suspended solids by more than 500,000 pounds per year and discharge of biochemical oxygen demand and nutrients by approximately 300,000 pounds per year.

> The term **waters of the United States** is defined at 40 CFR 122.2. It means:
>
> - Waters used for interstate or foreign commerce (e.g. Mississippi River).
> - All interstate waters, including interstate "wetlands."
> - Waters used for recreation by interstate or foreign travelers (for example a lake in one state that attracts fisherman from neighboring states).
> - Waters from which fish or shellfish are taken to sell in other states or countries.
> - Waters used for industrial purposes by industries involved in interstate commerce.
> - Tributaries and impoundments or dams of any waters described above.
> - Territorial seas.
> - Wetlands adjacent to any waters described above.
>
> Waters of the United States does not include:
>
> - Ponds or lagoons designed and constructed specifically for waste treatment systems.
> - Wetlands that were converted to cropland before December 23, 1985.
>
> These are only examples of kinds of waters considered waters of the United States. See the complete regulatory definition at 40 CFR 122.2 to see what other kinds of waters might be considered waters of the United States.
>
> The final regulation applies to CAAPs (that meet the production threshold) located in the territorial seas, contiguous zone, or ocean waters. Although EPA did not identify any existing facilities during the development of the regulation, net pens (or cages) operating in the contiguous zone or ocean waters would be subject to the regulation at this time. Future CAAPs (that meet the production threshold) in ocean waters or the contiguous zone are point sources subject to new source performance standards and NPDES permitting requirements.

The resulting improvements in water quality will create more opportunities for swimming and fishing, and reduce stress on ecosystems in those waters. They could also affect other aquatic environmental variables, such as primary production and populations or assemblages of native organisms in the receiving waters of regulated facilities.

Do other laws regulate CAAPs?

Although this guide explains what you need to do to comply with the federal CAAP regulation, your state, county, or town might have more requirements to address specific circumstances. Your permitting authority can set additional permit requirements if it finds them necessary. For example, they might set additional effluent limitations on a facility to ensure attainment of state water quality standards. State regulations must include federal requirements, but they can also be broader, stricter, or more specific. To learn about regulations in your state, contact your permitting authority. Appendix A contains a list of permitting authorities.

Your NPDES permit might include other federal requirements that apply to point source discharges (e.g., requirements under the Endangered Species Act, or resulting from the CWA Total Maximum Daily Load (TMDL) program). CAAPs might also be subject to federal requirements under the Animal and Plant Health Inspection Service (APHIS), the Spill Prevention, Containment, and Countermeasure (SPCC) regulations, CWA Section 403(c) Ocean Discharge Criteria, or the Federal Insecticide, Fungicide, and Rodenticide Act (FIFRA). New CAAPs may be subject to requirements resulting from implementation of the National Environmental Policy Act (NEPA). The following can provide technical assistance to make sure you are complying with all applicable requirements:

- Permitting authority/Departments of Environmental Protection – Appendix A
- National Association of State Aquaculture Coordinators (NASAC) http://www.marylandseafood.org/aquaculture/nasac.php – Appendix H
- State Sea Grant program – Appendix H
- Natural Resources Agencies Associated with Fisheries – Appendix B
- State Departments of Agriculture http://www2.nasda.org/NASDA
- USDA programs (Cooperative State Research, Education, and Extension Service (CSREES), Natural Resources Conservation Service (NRCS), others); see Appendix H for information about state cooperative extension service programs
  - General USDA programs – http://www.usda.gov
  - CSREES http://www.csrees.usda.gov
  - NRCS http://www.nrcs.usda.gov
- Other resources in your state

📖 *Final Preamble: Section XIII*

Table 1. Applicability of the CAAP ELGs to System Types

System Type or Subcategory	Annual Production (lb)	
	<100,000	*≥100,000*
Flow-through and Recirculating (Subpart A)	Not Applicable	Subject to: 451.3(a)–(d) 451.11(a)–(e) 451.12–14
Net pen (Subpart B)	Not Applicable	Subject to: 451.3(a)–(d) 451.21(a)–(h) 451.22–24

Table 2. Summary of Requirements for Flow-through and Recirculating Facilities

General Reporting Requirements	Reference
Drugs[1]	451.3(a)
1) Reporting of intention to use INADs where such use may lead to a discharge of the drug to waters of the U.S. • Provide the permitting authority with a written report, within 7 days of agreeing or signing up to participate in an INAD study • Identify the INAD to be used, method of use, the dosage, and the disease or condition the INAD is intended to treat	451.3(a)(1)
2) Oral reporting of INAD and extralabel drug use • Provide an oral report to the permitting authority as soon as possible, preferably in advance of application, but no later than 7 days after initiating use of the drug • Identify drugs used, method of application, and the reason for adding that drug	451.3(a)(2)
3) Written reporting of INAD and extralabel drug use • Provide a written report to the permitting authority within 30 days after initiating use of the drug • Identify the drug used and include the reason for treatment, date(s) and times(s) of the addition (including duration), method of application, and the amount added	451.3(a)(3)
Failure or Damage to the Structure of Aquatic Animal Containment System	451.3(b)
1) Specification of reportable damage and/or material discharge • The permitting authority may specify in the permit what constitutes reportable damage and/or material discharge of pollutants, based on consideration of production system type, sensitivity of the receiving waters, and other relevant factors	451.3(b)(1)
2) Oral reporting of structural failure or damage • Provide an oral report within 24 hours of the discovery of any reportable failure or damage that results in a material discharge of pollutants • Describe the cause of the failure or damage in the containment system • Identify materials that have been released to the environment as a result of the failure	451.3(b)(2)
3) Written reporting of structural failure or damage • Provide a written report within 7 days of discovery of the failure or damage • Document the cause of the failure or damage • Estimate the time elapsed until the failure or damage was repaired • Estimate materials released to the environment as a result of the failure or damage • Describe steps being taken to prevent a recurrence	451.3(b)(3)
Spills	451.3(c)
1) Oral reporting of spills of drugs, pesticides, and feed • Provide an oral report to the permitting authority within 24 hours of any spill of drugs, pesticides, and feed that results in a discharge to waters of the United States • Identify the material spilled and quantity	451.3(c)
2) Written reporting of spills of drugs, pesticides, and feed • Provide a written report to the permitting authority within 7 days of any spill of drugs, pesticides, and feed that results in a discharge to waters of the United States • Identify the material spilled and quantity	451.3(c)

[1] Reporting is not required for an INAD or extralabel drug use of a drug previously approved by FDA for a different aquatic animal species or diseases if the INAD or extralabel use is at or below the approved dosage and involves similar conditions of use.

Table 2. Summary of Requirements for Flow-through and Recirculating Facilities, Continued

Narrative Requirements		Reference
Best Management Practices Plan		451.3(d)
1) Development and maintenance of a BMP plan on site that describes how the permittee will achieve the following five requirements:		451.3(d)(1)
a) Solids control	• Employ efficient feed management and feeding strategies that limit feed input to the minimum amount reasonably necessary to achieve production goals and sustain targeted rates of aquatic animal growth in order to minimize potential discharges of uneaten feed and waste products to waters of the United States	451.11(a)
	• Identify and implement procedures for routine cleaning of rearing units and offline settling basins	
	• Identify procedures for inventorying, grading, and harvesting aquatic animals that minimize discharge of accumulated solids	
	• Remove and dispose of aquatic animal mortalities properly on a regular basis to prevent discharge to waters of the United States, except where authorized by the permitting authority in order to benefits the aquatic environment	
b) Material storage	• Ensure proper storage of drugs, pesticides, and feed in a manner designed to prevent spills that may result in the discharge of drugs, pesticides, or feed to waters of the United States	451.11(b)
	• Implement procedures for properly containing, cleaning, and disposing of any spilled materials	
c) Structural maintenance	• Routinely inspect production systems and wastewater treatment systems to identify and promptly repair damage	451.11(c)
	• Regularly conduct maintenance of production systems and wastewater treatment systems to ensure their proper function	
d) Record-keeping	• Maintain records for aquatic animal rearing units documenting feed amounts and estimates of the numbers and weights of aquatic animals in order to calculate representative feed conversion ratios	451.11(d)
	• Keep records documenting frequency of cleaning, inspections, maintenance, and repairs	
e) Training	• Train all relevant personnel in spill prevention and how to respond in the event of a spill to ensure proper clean-up and disposal of spilled materials	451.11(e)
	• Train personnel on proper operation and cleaning of production and wastewater treatment systems, including feeding procedures and proper use of equipment	
2) Make the plan available to the permitting authority upon request		451.3(d)(2)
3) Certify that a BMP plan has been developed		451.3(d)(3)

Table 3. Summary of Requirements for Net Pen Facilities

General Reporting Requirements		Reference
Drugs[2]		451.3(a)
1) Reporting of intention to use INADs where such use may lead to a discharge of the drug to waters of the U.S.	• Provide the permitting authority with a written report, within 7 days of agreeing or signing up to participate in an INAD study • Identify the INAD to be used, method of use, the dosage, and the disease or condition the INAD is intended to treat	451.3(a)(1)
2) Oral reporting of INAD and extralabel drug use	• Provide an oral report to the permitting authority as soon as possible, preferably in advance of application, but no later than 7 days after initiating use of the drug • Identify drugs used, method of application, and the reason for adding that drug	451.3(a)(2)
3) Written reporting of INAD and extralabel drug use	• Provide a written report to the permitting authority within 30 days after initiating use of the drug • Identify the drug used and include the reason for treatment, date(s) and times(s) of the addition (including duration), method of application, and the amount added	451.3(a)(3)
Failure or Damage to the Structure of Aquatic Animal Containment System		451.3(b)
1) Specification of reportable damage and/or material discharge	• The permitting authority may specify in the permit what constitutes reportable damage and/or material discharge of pollutants, based on consideration of production system type, sensitivity of the receiving waters, and other relevant factors	451.3(b)(1)
2) Oral reporting of structural failure or damage	• Provide an oral report within 24 hours of the discovery of any reportable failure or damage that results in a material discharge of pollutants • Describe the cause of the failure or damage in the containment system • Identify materials that have been released to the environment as a result of the failure	451.3(b)(2)
3) Written reporting of structural failure or damage	• Provide a written report within 7 days of discovery of the failure or damage • Document the cause of the failure or damage • Estimate the time elapsed until the failure or damage was repaired • Estimate materials released to the environment as a result of the failure or damage • Describe steps being taken to prevent a recurrence	451.3(b)(3)
Spills		451.3(c)
1) Oral reporting of spills of drugs, pesticides, and feed	• Provide an oral report to the permitting authority within 24 hours of any spill of drugs, pesticides, and feed that results in a discharge to waters of the United States • Identify the material spilled and quantity	451.3(c)
2) Written reporting of spills of drugs, pesticides, and feed	• Provide a written report to the permitting authority within 7 days of any spill of drugs, pesticides, and feed that results in a discharge to waters of the United States • Identify the material spilled and quantity	451.3(c)

[2] Reporting is not required for an INAD or extralabel drug use of a drug previously approved by FDA for a different aquatic animal species or diseases if the INAD or extralabel use is at or below the approved dosage and involves similar conditions of use.

Table 3. Summary of Requirements for Net Pen Facilities, Continued

Narrative Requirements		Reference
Best Management Practices Plan		451.3(d)
1) Develop and maintain a BMP plan on site that describes how the permittee will achieve the following seven requirements:		451.3(d)(1)
a) Feed management	• Employ efficient feed management and feeding strategies that limit feed input to the minimum amount reasonably necessary to achieve production goals and sustain targeted rates of aquatic animal growth	451.21(a)
	• Minimize accumulation of uneaten feed beneath the pens through active feed monitoring and management strategies approved by the permitting authority	
b) Waste collection and disposal	• Collect, return to shore, and properly dispose of all feed bags, packaging materials, waste rope, and netting	451.21(b)
c) Transport or harvest discharge	• Minimize any discharge associated with the transport or harvesting of aquatic animals (including blood, viscera, aquatic animal carcasses, or transport water containing blood)	451.21(c)
d) Carcass removal	• Remove and dispose of aquatic animal mortalities properly on a regular basis to prevent their discharge into the waters of the United States	451.21(d)
e) Materials storage	• Ensure proper storage of drugs, pesticides, and feed in a manner designed to prevent spills that may result in the discharge of drugs, pesticides, or feed into waters of the United States	451.21(e)
	• Implement procedures for properly containing, cleaning, and disposing of any spilled material	
f) Maintenance	• Inspect production systems on a routine basis in order to identify and promptly repair any damage	451.21(f)
	• Conduct regular maintenance on the production system in order to ensure its proper function	
g) Record-keeping	• Maintain records for aquatic animal net pens documenting the feed amounts and estimates of the numbers and weight of aquatic animals in order to calculate representative feed conversion ratios	451.21(g)
	• Keep records of net changes, inspections, and repairs	
h) Training	• Train all relevant personnel in spill prevention and how to respond to spills to ensure proper clean-up and disposal of spilled materials	451.21(h)
	• Train staff on proper operation and cleaning of production system, including feeding procedures and equipment	
2) Make the plan available to the permitting authority upon request		451.3(d)(2)
3) Certify that a BMP plan has been developed		451.3(d)(3)

Chapter 3: Does the CAAP Regulation Affect Me?

CAAP ELGs apply to owners and operators of CAAP facilities that meet certain conditions. If you produce more than 100,000 pounds annually, you may be subject to the ELGs.

Aquaculture facilities will fall into one of the following categories:

- No NPDES permit required
- Only NPDES permit required
- NPDES permit with ELGs requirements

If your aquatic animal operation is a CAAP under the NPDES regulations, you must apply for an NPDES permit. Refer to the next two sections in this chapter for additional information about which types of aquaculture facilities are required to apply for a permit (or renew their permit when their current permit expires if they are already permitted).

This chapter provides information about which operations are CAAPs and subject to NPDES permitting requirements, which are covered under the CAAP ELGs, what to do if you have more than one type of system at your facility, which facilities do not need an NPDES permit, how you know that your facility is not a CAAP, how a facility is defined, and what part of a facility is regulated.

What operations are CAAPs under the NPDES regulation?

EPA's existing NPDES regulations define when a hatchery, fish farm, or other facility is a CAAP facility and, therefore, a point source subject to the NPDES permit program. (See 40 CFR 122.24.) In defining CAAP facilities, the NPDES regulations distinguish between warm water and cold water species of fish and define a CAAP facility by, among other things, the size of the operation and frequency of discharge.

> *The criteria described in Appendix C of 40 CFR 122 are as follows. A hatchery, fish farm, or other facility is a CAAP facility if it grows, contains, or holds, aquatic animals in either of two categories: cold water species or warm water species.*
>
> *The cold water species category includes facilities where animals are produced in ponds, raceways, or other similar structures that discharge at least 30 days per year but does not include facilities that produce less than approximately 9,090 harvest weight kg (approximately 20,000 lb) of aquatic animals per year. It also does not include facilities that feed less than 2,272 kg (approximately 5,000 lb) of food during the calendar month of maximum feeding.*
>
> *The warm water species category includes facilities where animals are produced in ponds, raceways, or other similar structures that discharge at least 30 days per year. It does not include closed ponds that discharge only during periods of excess runoff or facilities that produce less than 45,454 harvest weight kg (approximately 100,000 lb) of aquatic animals per year.*

A facility is a CAAP facility if it meets the criteria in 40 CFR 122, Appendix C[1] or if it is designated as a CAAP facility by the

---

[1] 40 CFR 122, Appendix C is available in Appendix D of this document.

Director[2] on a case-by-case basis.

Most facilities falling under the definition of CAAP are either flow-through, recirculating or net pen systems. These systems discharge continuously or discharge 30 days or more per year as defined in 40 CFR 122.24 and are subject to permitting depending on the production level at the facility.

Most pond facilities do not require permits because ponds generally discharge fewer than 30 days per year and therefore generally are not CAAP facilities, unless designated by the Director.

Facilities meeting the NPDES definition of a CAAP will still be subject to the NPDES permit program, even if they are not subject to the requirements of the ELGs because their production levels are below 100,000 pounds per year.

Under 40 CFR 122, Appendix C:

"Cold water aquatic animals" include, but are not limited to, the Salmonidae family of fish; e.g., trout and salmon.

"Warm water aquatic animals" include, but are not limited to, the Ameiuride, Centrarchidae and Cyprinidae families of fish; e.g., respectively, catfish, sunfish and minnows.

Refer to Appendix L for a description of which systems the NPDES regulations cover.

---

[2] Director means the Regional Administrator or State Director, as the context requires, or an authorized representative. When there is no "approved state program," and there is an EPA administered program, "Director" means the Regional Administrator. When there is an approved state program, "Director" normally means the State Director. In some circumstances, EPA retains the authority to take certain actions even when there is an approved state program. ⬤ *Regulation: 40 CFR122.2*

What operations are covered under the CAAP ELGs?

The CAAP ELGs applies to direct dischargers of wastewater from these existing and new facilities (where production is defined as what leaves the facility):

- Facilities that produce at least 100,000 pounds a year in flow-through and recirculating systems that discharge wastewater at least 30 days a year (used primarily to raise trout, salmon, hybrid striped bass, and tilapia).
- Facilities that produce at least 100,000 pounds a year in net pens or submerged cage systems (used primarily to raise salmon).

Refer to Appendix L for a description of which systems are covered by the ELGs.

Figure 3.1. A flow-through system

What is the difference between NPDES and ELGs for CAAPs?

Any facility may be designated as a CAAP (if it meets the NPDES regulation requirements outright or if the Director designates it as a CAAP facility) and subject to NPDES permitting requirements.

However, if a CAAP facility is subject to ELGs requirements (i.e., recirculating, flow-through, or net pen systems that annually produce more than 100,000 pounds of aquatic animals) then the facility's NPDES permit will also contain ELGs requirements specific to the system types used to produce aquatic animals at that location. These are minimum requirements in the NPDES permit. A permit may contain additional more stringent limits required to ensure compliance with water quality standards.

What is considered a facility?

A facility is defined as all <u>contiguous</u> property and equipment owned, operated, or leased, or under control of the same person or entity. Each system owned, operated, leased, or under the control of the same person or entity that is <u>not contiguous</u> can and should be treated as separate facilities; the production threshold used in determining if a facility is a CAAP should also be applied separately.

Regulation: 40 CFR 451

What if I have more than one type of production system at my facility?

If you have more than one type of regulated system (flow-through, recirculating, or net pen) at your facility (and the combined annual production is 100,000 pounds or more), you must comply with the different requirements for each system type. For example, if you have a recirculating system and net pens at your facility, you will need to comply with the ELGs requirements for both recirculating systems and net pens. For more information about different system types and meeting the ELGs' production threshold, refer to the following examples.

The following are examples of combinations of system types that CAAP facilities may have, and whether the CAAP ELGs and NPDES requirements apply:

- *Recirculating – 25,000 pounds annually; Net pen – 80,000 pounds annually. (Both systems are regulated by the ELGs and NPDES requirements.)*

- *Flow-through – 75,000 pounds annually Recirculating – 50,000 pounds annually. (Both systems are regulated by NPDES requirements. If both are part of the same facility, the ELGs requirements also apply.)*

- *Flow-through – 25,000 pounds annually Recirculating – 50,000 pounds annually. (Neither is regulated by the ELGs; if growing coldwater species or determined to be a significant source of pollution to waters of the United States by the permitting authority, both are subject to NPDES requirements.)*

- *Flow-through – 125,000 pounds annually Pond – 15,000 pounds annually, discharging fewer than 30 days per year. (The flow-through system is regulated under the ELGs and NPDES; the pond is not regulated by the NPDES or ELGs regulations, unless the permitting authority determines that the pond is a significant source of pollution to waters of the United States.)*

- *Flow-through – 125,000 pounds annually Pond – 135,000 pounds annually and discharging more than 30 days per year. (The flow-through system is regulated under the ELGs and the pond is regulated by the NPDES regulations.)*

- *Net pen – 325,000 pounds annually Molluscan shellfish – 130,000 pounds annually. (The net pen is regulated by the ELGs and NPDES requirements; the molluscan shellfish system is not regulated by either unless the permitting authority determines it to be a significant source of pollution to waters of the United States.)*

Figure 3.2. A recirculating system

Figure 3.3. A net pen system

If you have other system types in addition to those subject to the ELGs, such as ponds or shellfish hatcheries, those system types are not subject to the ELGs (refer to page 3-5 of this chapter for a list of systems that are not subject to the ELGs). For example, if your facility has recirculating systems and ponds, only the recirculating systems are subject to the ELGs if the recirculating systems meet the annual production requirements of at least 100,000 pounds. The requirements for your recirculating system will appear in your NPDES permit. The ponds at your facility would not be subject to the ELGs requirements. However, you may need an NPDES permit if your ponds meet the definition of the cold water or warm water species category, where the ponds discharge at least 30 days per year (40 CFR 122, Appendix C) or if your pond is part of a facility that has been designated a CAAP facility.

If you are unsure which system types at your facility are subject to the ELGs requirements, contact your permitting authority.

Are there any aquaculture facilities that do not need an NPDES permit?

You do not need an NPDES permit if you are a facility that produces less than 9,090 harvest weight kilograms (approximately 20,000 pounds) per year of cold water species, if you feed less than 2,272 kilograms (approximately 5,000 pounds) of food during the calendar month of maximum feeding, or if you discharge less than 30 days per year (40 CFR122, Appendix C). However, you may need an NPDES permit if your facility is designated as a CAAP facility by the Director or if your state has more stringent requirements than EPA.

You do not need an NPDES permit if you are a facility that produces warm water species, using closed ponds that discharges only during periods of excess runoff, if you are a facility that produces less than 45,454 harvest weight kilograms (approximately 100,000 pounds) per year of warm water species, or if you discharge less than 30 days per year (40 CFR 122, Appendix C). However, you may need an NPDES permit if your facility is designated as a CAAP facility by the Director or if your state has more stringent requirements than EPA.

How do I know if I am not covered by these regulations?

In most cases, you are not covered by the NPDES or ELGs regulations if your production is less than the annual production thresholds covered by the regulations. However, if the Director designates your facility as a CAAP facility or if your state has more stringent requirements than EPA, you can be subject to NPDES permit requirements.

Systems not covered by the CAAP ELGs include:

- Closed pond systems (may be covered by NPDES if discharges occur more than 30 days per year or if designated as a CAAP facility by the Director)
- Molluscan shellfish (including nurseries)
- Shrimp ponds
- Crawfish production
- Alligator production
- Aquaria
- Net pens rearing native species released after a growing period of no longer than 4 months to supplement commercial and sport fisheries.

What if I discharge to a POTW?

The CAAP ELGs do not establish national pretreatment standards for facilities that meet the criteria for a CAAP facility (as defined in 40 CFR 122.24 and Appendix C of 40 CFR 122) and are indirect dischargers (i.e., facilities that discharge to a publicly owned treatment

> *An indirect discharger is a facility that discharges or may discharge wastewaters into a publicly-owned treatment works.*

works (POTW)). However, you may be subject to local limit requirements.

National pretreatment standards are established for pollutants in wastewater from indirect dischargers that may pass through, interfere with, or are otherwise incompatible with POTW operations. Generally, pretreatment standards are designed to ensure that wastewaters from direct and indirect industrial dischargers are subject to similar levels of treatment. POTWs are required to implement local treatment limits applicable to their indirect dischargers to satisfy any local requirements. You should communicate with your POTW operator to determine any local pretreatment standards that apply to your facility.

> *A Publicly Owned Treatment Works (POTW) is a treatment works (as defined by section 212 of the CWA), which is owned by a state or municipality (as defined by section 502(4) of the CWA). This definition includes any devices and systems used in the storage, treatment, recycling, and reclamation of municipal sewage or industrial wastes of a liquid nature. It also includes sewers, pipes, and other conveyances, only if they convey wastewater to a POTW. The term also means the municipality, as defined in section 502(4) of the CWA, that has jurisdiction over the indirect discharges to and the discharges from such a treatment works.*

If you discharge to a POTW, contact your permitting authority for more details.

What part of my CAAP is regulated?

The CAAP regulation applies to the production areas of your facility, including:

- Areas where you might grow, maintain, or contain aquatic animals (e.g., raceways, tanks, or net pens).
- Areas where you might store raw materials (e.g., feed silos and storage areas designated for feed or drugs).
- Areas where you might contain wastes (e.g., sedimentation basins, quiescent zones, and settling ponds).
- Source water and wastewater conveyance systems (e.g., tailraces and headraces).

No specific guidance for land application of waste was developed for the CAAP ELGs. If a facility is doing land application, a good source of information regarding land application is EPA's *Producers' Compliance Guide for CAFOs*, available at http://www.epa.gov/npdes/pubs/cafo_prod_guide_cover_and_contents.pdf.

Figure 3.4. A feed storage area

Chapter 4: What Do I Need to Know About NPDES Permits?

How do I apply for an NPDES permit?

To apply for a permit, you first need to acquire the application forms. You can get the forms you need to apply for an NPDES permit from your permitting authority. Some states have made the forms available on their websites. Check with your permitting authority to make sure you are using the correct forms.

> *An applicant should consider requesting a pre-application meeting to resolve any questions and to seek guidance from the agency and or permit writer. It may be advantageous for the applicant, permit writer, and aquaculture extension specialist to hold joint discussion to develop BMP components appropriate for the specific facility, species, production system, and location under consideration.*

You next need to determine what type of permit you will be applying for. Under the federal NPDES regulations, there are two types of permits—general permits and individual permits. Each permitting authority adopts its own rules about what types of permits operations need, so you should contact your permitting authority for more information.

After you determine what type of NPDES permit to apply for, you need to complete the application forms and submit the required information. Refer to page 4-3 for a discussion of the information you must include in your permit application.

Depending on your specific facility (existing, new, etc.), you must apply for a renewal of your current permit or a new permit by the required deadlines. Refer to page 4-3 for more information about deadlines for applying for NPDES permits.

After you acquire an NPDES permit, you must have the permit in effect for your operation as long as it is an operating CAAP. Refer to page 4-6 for a discussion of situations where you can discontinue your NPDES coverage. Page 4-7 contains information about what to do if significant changes occur at your operation.

What is an NPDES general permit?

An NPDES general permit has one set of requirements for a group of similar types of facilities. For example, all CAAP facilities in a particular area, such as an entire state or a watershed within the state, might be covered under one general permit. The permitting authority sets the permit conditions, issues a draft permit, and requests comments from the public. The permitting authority may make changes to the draft permit based on the public comments and then issues the final permit.

The general permit specifies what kinds of operations can be covered. Owners and operators of eligible operations may then apply for coverage under the general permit. Contact your permitting authority to see if your facility is eligible to be covered under an existing general permit.

Operators of CAAP facilities that are eligible for coverage under a general permit may notify the permitting authority that they want to be covered by submitting a Notice of Intent (NOI). If an NPDES general permit is available in your state and your operation meets the eligibility requirements, you must fill out an NOI and submit it to your permitting authority to apply for coverage under the general permit. The general permit will tell you how to apply for coverage, the deadline for applying, and when your coverage will become effective.

Some general permits specify that coverage is automatic unless notified by the permitting authority. Coverage under other general permits does not begin until receipt of notification of applicability by the permitting authority. If coverage is automatic, EPA recommends that the general permit specify that the facility is authorized to discharge in accordance with the permit after a specified waiting period of, for example, 30 days. Having a specified waiting period or coverage only upon receipt of a notification of applicability will allow the permitting authority the opportunity to provide for meaningful public involvement after NOIs are submitted.

States that have developed general aquaculture NPDES permits include the following:

- *EPA Region 10 – General NPDES Permit for Aquaculture Facilities in Idaho and Associated, On-site Fish Processors (ID-G13-0000)*
- *Maine – General Permit for Atlantic Salmon Aquaculture (MEG130000)*
- *North Carolina – General Permit No. NCG530000*
- *Washington – Upland Fin-Fish Hatching and Rearing General NPDES Permit*

What is an NPDES individual permit?

An NPDES individual permit contains requirements designed specifically for one CAAP facility. You must apply for an

NPDES individual permit if any of the following are true:

- A general NPDES permit is not available.
- Your CAAP facility is not eligible to be covered under the general NPDES permit.
- You want an individual NPDES permit.
- Your permitting authority requires you to apply for an individual permit.

The permitting authority may also require any discharger currently covered by a general permit to apply for and obtain an individual NPDES permit.

To apply for an individual permit, you must fill out either NPDES Forms 1 and 2B (available in Appendix K) or similar forms required by your state. You should contact your permitting authority for the proper forms. Forms 1 and 2B may be downloaded from EPA's website at: http://cfpub2.epa.gov/npdes/doctype.cfm?sort=name&program_id=45&document_type_id=8. Your state permitting authority may also provide the necessary application forms on their websites. Check with your permitting authority to be sure you submit the correct forms.

You must complete the forms and submit them to your permitting authority. When your permitting authority receives your permit application, it will use the information you submitted to draft a permit for your operation. Your permitting authority will base your permit requirements on the unique conditions at your operation. A collaborative effort between the farmer, permit writer, and an aquaculture extension specialist may be helpful. After a public comment period on the draft permit, your permitting authority will modify the draft, if necessary, and then issue your final NPDES individual permit.

What information do I have to include in my NOI or permit application?

When you apply for an individual NPDES permit, you must give the following information to your permitting authority, as part of Form 2B (much of the same information may be required as part of an NOI for coverage under a general permit):

- Contact information for the owner or operator of the facility.
- If the facility is existing or proposed.
- The location and mailing address of your facility.
- The latitude and longitude of the entrance to your facility's production area.

  > Check EPA's website at: http://cfpub.epa.gov/npdes/stormwater/latlong.cfm to find out how to determine the latitude and longitude and where to get a topographic map for your location.

- A topographic map of the area where your facility is located, with the location of the production area specifically marked.
- The outfall number and flow for each outfall from the facility.
- The total number and size of ponds, raceways, tanks, other rearing units, and similar structures at your facility.
- The name of the receiving water.
- The source of water used in your facility.
- The species (cold water and warm water) of fish or aquatic animals held at your facility. For each species, you will need to provide the total weight produced by your facility per year in pounds of harvestable weight (harvestable weight = gross production), as well as the maximum weight present at any one time.
- The total pounds of food fed during the calendar month of maximum feeding.
- The treatment systems and practices you use for wastewater.

Your permitting authority may require more information than what is listed above when you apply for a permit. Check with your permitting authority to make sure you are submitting the correct information.

📖 *Regulations: 40 CFR 122.21 and 122.28*

When do I have to get an NPDES permit?

Your permit application deadline depends on whether your operation is an existing CAAP facility, a new discharger, or a new source. Each category has a different deadline for applying for an NPDES permit. Read the descriptions below to determine when you must apply for an NPDES permit.

- If you are an *existing CAAP facility* (already have an NPDES permit), you must apply to renew your NPDES permit at least 180 days before it expires. Refer to the section "Existing CAAP Facilities" below for additional information.
- If you are a *new discharger*, you must apply for an NPDES permit at least 180 days before you plan to begin discharging from the CAAP facility. Refer to the section "New Dischargers" below for additional information.
- If you are a *new source*, you must apply for an NPDES permit at least 180 days before you plan to begin discharging from the CAAP facility.

Refer to the section "New Sources" below for additional information.

You are responsible for applying for NPDES permit coverage for your facility. The federal regulations do not require your permitting authority to notify you that you must apply. For an individual permit, the permitting authority issues a permit after it receives a complete and accurate permit application from the facility seeking coverage. For a general permit, the permitting authority issues the general permit, and operators then submit their NOIs to be covered under the permit. In both instances, the permitting authority is required to provide public notification that a permit has been drafted. In addition, although permitting authorities are not required to do so, many are likely to conduct outreach to communicate who must obtain a permit and how to do so. Ultimately, however, the responsibility to seek permit coverage lies with the aquaculture facility. Your failure to meet the permitting deadlines described below could result in liability under the Clean Water Act, which may result in penalties.

Existing CAAP Facilities

Existing CAAP facilities are operations that are already permitted under 40 CFR 122.24. If you operate a CAAP facility that is already permitted, you already have an NPDES permit. You will have to reapply for a new permit at least 180 days before your existing permit expires. When the permitting authority renews your permit, your permit will include the ELGs requirements *if* you meet the production and system type applicability requirements of the ELGs. If you *do not* meet the production and system type applicability requirements of the ELGs, recall that other NPDES requirements may apply to your facility.

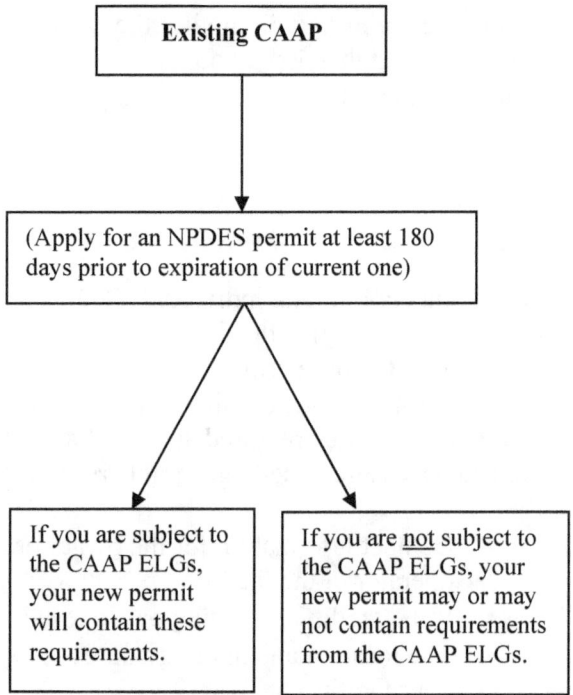

New Dischargers

New dischargers are operations that are defined as CAAPs after the effective date of the rule (September 22, 2004), but are not new sources. A general definition of new discharger is found at 40 CFR 122.2.

One example of a new discharger is a facility that is newly constructed, will meet the production threshold for a CAAP (thus requiring an NPDES permit), but will not meet the thresholds at which ELGs, and therefore New Source Performance Standards, would apply. As a new discharger, you must apply for a permit at least 180 before you plan to commence discharging unless permission for a later date is granted by the permitting authority.

It also is possible for a facility already constructed and discharging to be considered a new discharger under the NPDES program. For example, if your facility previously was not defined as a

CAAP but plans to increase the number of aquatic animals produced so you exceed the production threshold for definition as a CAAP facility, you would then be considered a new discharger for purposes of NPDES permitting. In this case, unless a later date is granted by the permitting authority, you must apply for an NPDES permit at least 180 days before you increase production and, therefore begin discharging as a point source subject to NPDES requirements.

New Sources

A CAAP facility is a new source if construction of the facility began after September 22, 2004 and the CAAP ELGs apply to the facility. Under the CWA, construction refers to the construction of any building, structure, or facility and to the installation of equipment. Construction commences if an entity either undertakes or begins certain work as part of a continuous on-site construction program, or enters into contractual obligations to purchase facilities or equipment. If construction occurs after the new source date, the facility will be considered a new source if it meets any of the following criteria:

- It is constructed at a site at which no other source is located; or
- It totally replaces the process or production equipment that causes the discharge of pollutants at an existing source; or
- Its processes are substantially independent of an existing source at the same site. To determine whether the processes are substantially independent, factors such as the extent to which the new facility is integrated with the existing plant; and the extent to which the new facility is engaged in the same

general type of activity as the existing source should be considered. 40 CFR 122.29(b)(1), 40 CFR 403.3 (k)(1).

Construction at land-based sites such as flow through and recirculating systems occurs when ground is broken, new equipment is delivered, or other significant changes occur. For net pen and cage operations, construction at the facility is generally considered the start of the timeframe for a new source. Construction is considered to include when some activity associated with site preparation or construction of the pens in the water occurs. For example:

- Some activity takes place at the site bottom or surface in preparation for placement of a net pen or cage.
- If no site bottom or surface preparation is necessary, net pens or cages are placed in the water.

The date construction at the facility begins compared to the effective date of the rule is important. A new source determination for aquatic animal production facilities for water-based systems such as net pens will be based on the date at which construction commences. Based on individual circumstances, the date construction begins may be the date the nets, cages, or structure, are placed in the water, date the nets/cages were purchased, or the date a binding contractual agreement takes place.

New sources will need to comply with the NSPS and limitations of the CAAP rule at the time such sources commence discharging CAAP process wastewater.

If you own or operate a new source CAAP facility, you must apply for a permit at least 180 days before you plan to begin discharging from the CAAP facility. New source permitting is subject to National Environmental Policy (NEPA) review.

When will my NPDES permit expire?

Individual NPDES permits are usually issued for 5-year terms and are reissued every 5 years. You should check the expiration date of your permit.

General NPDES permits are also usually issued for 5-year terms. Because a general NPDES permit is created for multiple permittees, however, it could have been issued several years before you submitted your NOI. If this is the case, the general NPDES permit might expire less than 5 years after you submit your NOI.

To reapply for a permit when it is due to expire, you must submit a complete and accurate new application form (for an individual permit) or new NOI (to be covered under a general permit) 180 days before your permit's expiration date. For *EPA issued permits*, if you have met this deadline and your permitting authority fails to reissue your NPDES permit before the expiration date, the conditions of your current NPDES permit will remain in effect until the permitting authority acts on your complete and accurate new application (40 CFR 122.6). Although many states have automatic continuation, *state issued permits* are subject to state law.

Some permitting authorities might have other deadlines or procedures for reissuing CAAP NPDES permits. For example, some general permits are automatically continued without a facility submitting a new NOI. Check the reapplication procedures specified in your permit, and contact your permitting authority to find out exactly what you must do to get a new permit when your current permit is due to expire.

How long should I keep my NPDES permit?

You must have an NPDES permit in effect for your operation as long as it is an operating CAAP. There are a few situations in which you can discontinue your NPDES coverage:

- You close your operation.
- You *permanently change* your operation so that it no longer is a CAAP (under the NPDES program).

Under all circumstances, you must have an NPDES permit in effect until you properly dispose of process wastewater that was generated at the CAAP facility and solids collected in a settling basin or held in a storage tank so that your operation no longer has a potential to discharge to waters of the United States. If your operation still has a potential to discharge when your permit is due to expire, you must reapply for a permit. Once you have properly disposed of the collected solids, and process wastewater so that there is no longer a potential to discharge, you may ask your permitting authority to terminate your permit. Contact your permitting authority to find out more about how to terminate your permit. (You can find contact information for your permitting authority in Appendix A of this guide.)

If you make short-term (1-2 years) changes to your operation that reduce annual production so you no longer meet the definition of a CAAP facility, you may request changes be made to your permit. However, remember that the permitting authority can reevaluate your operation and add requirements at that time. Contact your permitting authority if you have any questions.

What if I make significant changes to my operation while I have an NPDES permit?

If you make significant changes at your operation, under NPDES regulations and the terms of your permit, you must contact your permitting authority and report these changes. Examples of significant changes include increasing production levels (e.g., increasing annual production from 50,000 pounds to 175,000 pounds), changes to structures (e.g., removing quiescent zones), or changes to facility configuration (e.g., adding 10 raceways to a facility).

Chapter 5: What Requirements Will My NPDES Permit Contain?

Your NPDES permit will state what you have to do to comply. Certain minimum requirements must be in every NPDES CAAP permit, and this guide describes those minimum requirements. Your permitting authority may include more than the minimum requirements in your NPDES permit. You should read your permit carefully to find out exactly what you have to do at your CAAP facility. Your NPDES permit will include the following requirements:

- Effluent limitations, if applicable
- Special conditions
- Standard conditions
- Monitoring, record-keeping, and reporting requirements

▭ *Regulation: 40 CFR 122.41*

> *Remember to read your permit and check with your permitting authority to find out exactly what your permit requires. This guide describes the minimum requirements established by the federal CAAP regulations. Your permit might require you to do more than the minimum requirements described here, for example, to meet your state's or tribe's water quality standards or to comply with CAAP requirements specific to your state. See the appendix to find out how to contact your permitting authority.*

What are the elements of an NPDES permit for a CAAP facility?

The elements of an NPDES permit for a CAAP are the same as those issued to other point sources. These elements consist of a cover page, effluent limitations, monitoring and reporting requirements, record keeping requirements, special conditions, and standard conditions.

Cover page – serves as the legal notice of the applicability of the permit, provides the authority under which it is issued, and contains appropriate dates and signature(s).

Effluent limitations and standards – serves as the primary mechanism for controlling discharges of pollutants to receiving waters (e.g., the specific narrative or numeric limitations applied to the facility and the point of application of these limits).

Monitoring and reporting requirements – identifies all of the specific conditions related to the types of monitoring to be performed, the frequencies for collecting samples or data, and how to record, maintain, and transmit the data and information to the permitting authority.

Record-keeping requirements – specifies the types of records to be kept on-site at the permitted facility (e.g., inspection and monitoring records).

Special conditions – in NPDES permits for CAAPs, special conditions may be included, as determined necessary by the permitting authority.

Standard conditions – conditions that apply to all NPDES permits, such as the requirement to properly operate and maintain all facilities and systems of treatment and control, as specified in 40 CFR 122.41.

For additional details on the elements of an NPDES permit, refer to the U.S. EPA NPDES Permit Writers' Manual (EPA-833-B-96-003).

What effluent limitations will be included in my NPDES permit?

Your permit will contain technology-based effluent limitations (based on the amount of pollutant reduction that can be achieved by applying pollution control technologies or practices), water quality-based effluent limitations (based on the water quality standards for and the condition of the receiving water body), or both. It might also contain additional BMPs, as needed.

> *The technology-based limitations or requirements in a CAAP permit will be based on the ELG, for pollutants covered by the ELGs. The permit writers using best professional judgment (BPJ) may develop so called BPJ limits.*
>
> *A water quality-based effluent limitation is designed to protect the quality of the receiving water by ensuring that state or tribal water quality standards are met. In cases where a technology-based requirement does not sufficiently protect water quality, the permit must include appropriate water quality-based limits.*
>
> *For example, a technology-based standard for a CAAP facility might require the development of a facility BMP plan that includes controlling the discharge of solids. At some facilities, additional controls may be required to further reduce the discharge of phosphorus because of excessive nutrient loading in the receiving waterbody that may result in exceeding water quality standards. For these facilities, a water quality-based effluent limitation in the form of numeric phosphorus limits, such as seasonally-adjusted monthly maximum loads for total phosphorus, may be included in the permit to reduce the discharge of phosphorus and ensure that water quality standards are met.*

Effluent limitations for flow-through and recirculating facilities

As explained in detail in Chapter 2 (Table 2), the ELGs contain specific reporting

activities and narrative requirements (i.e, management practices) for flow-through and recirculating facilities that produce at least 100,000 pounds of aquatic animals annually.

The CAAP ELGs contain general reporting requirements for the use of certain types of drugs. All CAAP facilities that are subject to 40 CFR 451

> *Reporting is not required for an INAD or extralabel drug use that has been previously approved by FDA for a different aquatic animal species or diseases if the INAD or extralabel use is at or below the approved dosage and involves similar conditions of use.*

must notify the permitting authority of the use of any investigational new animal drug (INAD) and any extralabel drug use where the use may lead to a discharge to waters of the United States. The ELGs also contain general reporting requirements for failure in or damage to the structure of an aquatic animal containment system, resulting in an unanticipated material discharge of pollutant to waters of the United States.

The CAAP ELGs contain narrative requirements for management practices for flow-through and recirculating facilities. Under these requirements, you must develop and maintain a BMP plan on site that

> *Facilities should note that the management practices are general (e.g., solids control) and a facility may choose how to achieve the management practice. For example, solids control can be achieved through feed management or solids disposal. EPA does not specify what a facility must do to achieve solids control.*

describes how you will manage the following:

- Solids control
- Material storage
- Structural maintenance

- Record-keeping
- Training

📖 *Final Preamble: Section VIII.B*

Effluent limitations for net pen facilities

As explained in detail in Table 3 of Chapter 2, the ELGs require management practices and record-keeping activities for net pen facilities that produce at least 100,000 pounds of aquatic animals annually.

> Again, reporting is not required for an INAD or extralabel drug use that has been previously approved by FDA for a different aquatic animal species or diseases if the INAD or extralabel use is at or below the approved dosage and involves similar conditions of use.

The ELGs contain general reporting requirements for the use of certain types of drugs. All CAAP facilities that are subject to 40 CFR 451 must notify the permitting authority of the use of any INAD and any extralabel drug use where the use may lead to a discharge to waters of the United States. The ELGs also contain general reporting requirements for failure in or damage to the structure of an aquatic animal containment system, resulting in an unanticipated material discharge of pollutants to waters of the United States.

The ELGs contain narrative management practice requirements for net pen facilities. Under these requirements, you must develop and maintain a BMP plan on site that describes how you will manage the following:

- Feed management
- Waste collection and disposal
- Transport or harvest discharge
- Carcass removal

- Material storage
- Maintenance
- Record-keeping
- Training

📖 *Final Preamble: Section VIII.C*

What are special conditions?

Some NPDES permits contain special conditions that supplement the effluent limitations. Special conditions address unique conditions at an operation. Typical special conditions include, for example, BMPs, special monitoring studies, and stream surveys.

What special conditions will be included in my CAAP NPDES permit?

The ELGs do not impose any special conditions in your CAAP NPDES permit. However, your permit may contain special conditions to address local concerns. For example, where authorized, net pen facilities may be required to perform regular benthic monitoring to ensure that solids are not accumulating under the net pens and causing harm to benthic communities. Other additional requirements may address spills (e.g., petroleum), protection for endangered species and migratory birds, employee training, and groundwater monitoring or the use of liners in areas where there is the potential for a discharge to groundwater that has a direct hydrologic connection to waters of the United States. In addition, states concerned with groundwaters as waters of the state may require monitoring, liners, or other requirements based on appropriate state authority.

What are the standard conditions of all NPDES permits?

Most NPDES permits contain standard conditions, which include definitions, testing procedures, record-keeping requirements, penalties for noncompliance, and your responsibilities as an NPDES permit holder. These responsibilities include, for example, complying with your permit, meeting deadlines for reapplying when your permit is due to expire, properly operating and maintaining your facility, and letting the permitting authority inspect your operation. The standard conditions also require you to notify your permitting authority if certain things happen at your operation (e.g., a significant increase in annual production or an upset occurs). See Chapter 4 of this guide for additional information. Carefully read the standard conditions section of your NPDES permit, and contact your permitting authority if you have any questions.

📖 *Regulation: 40 CFR. 122.41*

What records do I have to keep?

Your NPDES permit will require you to keep certain records to show that you are complying with the terms of the permit. You must keep all the records on-site at your operation for 5 years and you must provide them to the permitting authority upon request.

What are the record-keeping requirements for all CAAPs, under the ELGs?

If you own or operate a flow-through, recirculating, or net pen system that produces 100,000 pounds or more each year, you must keep at least the following records that document:

- Feed amounts and estimates of the numbers and weights of aquatic animals in order to calculate representative feed conversion ratios.

- Frequency of cleaning, inspections, maintenance, and repairs.

- Net pen changes, inspections, and repairs.

Refer to Appendix R for a checklist of record-keeping requirements. Chapter 12 of this document ("Record-keeping for Flow-through, Recirculating, and Net Pen Facilities") provides a more detailed discussion of record-keeping and refers to example forms in the appendices that may be used for record-keeping.

📖 *Final Preamble: VIII.E*

What monitoring do I have to perform under my NPDES permit?

The monitoring that your permitting authority may require as part of your permit depends on other conditions in your permit. If you are subject to ELGs, there are some associated monitoring requirements, as discussed below. If your permit includes numeric effluent limitations, you will be required to monitor to demonstrate compliance with those limitations. Your permitting authority may also require monitoring to characterize your discharge even when your permit does not include numeric effluent limitations. Look carefully at your permit, particularly the effluent limitations section, special conditions, and monitoring requirements section, to find out what monitoring you have to perform.

What monitoring do I have to perform under the ELGs?

Under the ELGs, you must monitor your production systems and wastewater treatment systems for damage to structural components. More specifically, you must do the following:

- Routinely inspect production systems and wastewater treatment systems to identify and promptly repair damage. FT RAS

- Routinely inspect production systems to identify and promptly repair damage. NBT

Example forms for tracking inspections are available in Appendix P.

Final Preamble: VIII.D

What do I have to report to the permitting authority?

Your permit may require you to submit certain reports to your permitting authority, such as a monitoring report; an annual report; or special reports of discharges, changes to your operation, and other information, such as the use of certain drugs through INADs or extralabel prescriptions. Read your permit carefully, and contact your permitting authority to find out exactly what you must report.

Chapter 12 of this document provides a list of example forms and logs available in the appendices that may be used to report required information to your permitting authority or for record-keeping.

Final Preamble: VIII.D

What else do I have to report?

The standard conditions that apply to all NPDES permits (refer to Chapter 4 of this guide for additional information) also include the following reporting requirements:

- *Duty to provide information.* You must provide any information your permitting authority needs to find out if you are complying with your NPDES permit or to make changes to your permit.
- *Signatory and certification requirements.* Any applications, reports, or information you submit must be signed and certified. The certification must state that all the information you submit is true and complete to the best of your knowledge. There might be penalties if you knowingly submit false information.
- *Planned changes.* If you plan to make any changes to your CAAP facility that will affect your ability to comply with your NPDES permit, you have to notify your permitting authority as soon as possible.
- *Anticipated noncompliance.* You must notify your permitting authority if you know that something is going to happen at your facility that would cause you to be out of compliance with your NPDES permit. Failing to do so could result in penalties.
- *Twenty-four-hour reporting.* If you have a discharge (or other noncompliance event) at your CAAP facility that could endanger human health or the environment, you must report it orally within 24 hours. Within 5 days, you must submit a written statement describing the discharge or noncompliance. Your

description must include what caused the discharge, when it started, how long it lasted, what you did to stop the discharge, and how you will prevent the problem in the future.

- *Other noncompliance.* You must report all instances of noncompliance. Each report must contain the information described above for twenty-four hour reporting.
- *Other information.* If you find out that you failed to submit any important facts in your application, or that you submitted incorrect information in your application or other reports, you must submit the correct information right away.

Regulation: 40 CFR 122.41 (h), (k), and (l)(1), (2), (6), (7), and (8)

What other requirements might my permit contain?

Your NPDES permit might also contain requirements to address other considerations, such as considerations to implement requirements under the CWA Total Maximum Daily Load (TMDL) programs. Check with your permitting authority if you have any questions about these other requirements.

A TMDL is a calculation of the greatest amount of a pollutant that a waterbody can receive without exceeding water quality standards. It is the sum of the allowable loads of a single pollutant from all contributing point and nonpoint sources. The calculation must include a margin of safety to ensure that the waterbody can be used for the purposes the state has designated. The calculation must also account for seasonal variation in water quality. Additional information about TMDLs is available from EPA's TMDL website at http://www.epa.gov/owow/tmdl.

Chapter 6: General Reporting Requirements for Flow-through, Recirculating, and Net Pen Facilities [FT] [RAS] [NET]

EPA established general reporting requirements for the use of certain types of drugs (i.e., Investigational New Animal Drugs (INADs), extralabel prescriptions). EPA also established general reporting requirements for failure in or damage to the structure of an aquatic animal containment system, resulting in an unanticipated material discharge of pollutant to waters of the United States.

What is an INAD drug?

An INAD is a drug for which there is a valid exemption in effect under 512(j) of the Federal Food, Drug, and Cosmetic Act, 21 U.S.C. 360b(j). More specifically, INADs are those drugs for which FDA has authorized use on a case-by-case basis to allow a way of gathering data for the approval process. Quantities and conditions of use are specified. FDA, however, sometimes relies on the NPDES permitting process to establish limitations on pollutant discharges to prevent environmental harm. Most NPDES permits, which mention drugs and pesticides, to date have required only reporting of the use of drugs and pesticides.

FDA may grant INAD exemptions from approved use for establishing data to base drug approval. Through the investigative approval process, the sponsor agrees to conduct laboratory and field tests with the drug under the conditions and on the animals proposed for approval. These data are collected in the INAD and eventually submitted to a new animal drug application (NADA) to form the basis for the Center for Veterinary Medicine's (CVM's) approval or disapproval of the drug. Data collection for

the drug approval includes data on the observed or anticipated environmental effects associated with the drug's use. In the case of drugs used on aquatic animals the most significant environmental effect associated with the drug's usage is the effect on the aquatic environment.

What is extralabel use of a drug?

Extralabel use is when a drug is not used according to label requirements. Extralabel drug use is restricted to use of approved animal and human drugs by, or on the order of, a licensed veterinarian and must be within the context of a valid veterinarian-patient relationship.

An example of an extralabel use is injecting erythromycin into adult fish to treat bacterial infections. Since there are no current labeled uses of erythromycin in aquatic animals, this use would require a veterinarian to provide an extralabel prescription. Note, although erythromycin is under an INAD exemption to control bacterial kidney disease in salmonids, any uses other than those associated with the INAD study are only allowed as an extralabel use.

> *Your veterinarian prescribes oxytetracycline (Terramycin) medicated feed at a dose of 3.0 g of drug per 100 lb of feed for your yellow perch (grown in a flow-though system) that have been diagnosed with* Aeromonas liquefaciens. *Since Terramycin is approved for as a feed additive to treat salmonids and catfish for* Aeromonas liquefaciens, *when used according to the veterinarian's prescribed instructions, you do not have to report the use. Remember, you still are required to keep records of the treatment conditions as a requirement of the extralabel provisions developed by FDA.*

What am I required to do if I use an INAD drug?

Unless you are exempt, under the general reporting requirements, you must first report your intention to use INAD(s). You must provide a written

> **INAD Reporting Exemption**
>
> **Remember:** you do not need to report an INAD use of a drug previously approved by FDA for a different aquatic animal species or disease if:
>
> - The dosage of the drug is used at less than or equal to the approved dosage **and**
> - The use is done under similar conditions.

report to the permitting authority of an INAD's impending use within 7 days of agreeing or signing up to participate in an INAD study. The written report must identify the:

- INAD to be used.
- Method of application.
- Dosage.
- Disease or condition the INAD is intended to treat.

Second, you must provide the permitting authority with an oral report that you are using INAD(s). You must provide an oral report to the permitting authority as soon as possible (preferably in advance of use), but no later than 7 days after initiating use of the drug. The oral report must identify the:

- Drugs used.
- Method of application.
- Reason for using the drug.

Finally, you must also provide a written report to your permitting authority that you are using INAD(s). You must provide a written report to the permitting authority

within 30 days after initiating use of the drug. The written report must identify the:

- Drugs used.
- Reason for treatment.
- Date(s) and time(s) of the addition (including duration).
- Method of application.
- Amount added.

> If using an **INAD drug**, you must provide the following to your permitting authority:
>
> 1. A written report within 7 days of agreeing to use an INAD.
> 2. An oral report no later than 7 days of initiating use of the drug.
> 3. A written report within 30 days after initiating use of the drug.

Refer to Appendix M for example forms that may be used to submit this information to your permitting authority.

📖 *Regulation: 40 CFR 451.3*

What am I required to do if there is extralabel drug use at my CAAP facility?

Unless you are exempt, you must provide the permitting authority with an oral report of extralabel drug use. You must provide an oral report to the

> **Extralabel Drug Use Reporting Exemption**
>
> **Remember:** you do not need to report an extralabel use of a drug previously approved by FDA for a different aquatic animal species or disease if:
>
> - The dosage of the drug is used at less than or equal to the approved dosage **and**
> - The use is done under similar conditions

permitting authority as soon as possible (preferably in advance of use), but no later than 7 days after initiating use of the drug. The oral report must identify the:

- Drugs used.
- Method of application.
- Reason for using the drug.

You must also provide a written report to your permitting authority of extralabel drug use. You must provide a written report to the permitting authority within 30 days after initiating use of the drug. The written report must identify the:

- Drugs used.
- Reason for treatment.
- Date(s) and time(s) of the addition (including duration).
- Method of application.
- Amount added.

With **extralabel drug use**, you must provide the following to your permitting authority:

1. An oral report no later than 7 days of initiating use of the drug.
2. A written report within 30 days after initiating use of the drug.

Refer to Appendix M for example forms that may be used to submit this information to your permitting authority.

📖 *Regulation: 40 CFR 451.3*

What am I required to do if there is a failure in, or damage to, the structure of an aquatic animal containment system?

You will need to notify your permitting authority if:

- There is any failure in, or damage to, the structure of an aquatic animal

containment system resulting in an unanticipated material discharge of pollutants to waters of the United States **and/or**
- If there is a spill of drugs, pesticides, or feed that results in a discharge to waters of the United States.

Upon discovery of a **structural failure or damage to a containment system or spill of drugs, pesticides, or feed**, you must provide the following to your permitting authority:

1. An oral report within 24 hours of discovery.
2. A written report within 7 days of discovery.

The permitting authority may specify in the permit what constitutes reportable damage and/or material discharge of pollutants, based on consideration of production system type, sensitivity of the receiving waters, and other relevant factors.

You must provide an oral report to your permitting authority within 24 hours of the discovery of any reportable failure or damage that results in a material discharge of pollutants. This report must:

- Describe the cause of the failure or damage in the containment system.
- Identify materials that have been released to the environment as a result of the failure.

You must also provide a written report to your permitting authority within 7 days structural failure or damage. This report must:

- Document the cause of the failure or damage.

- Estimate the time elapsed until the failure or damage was repaired.
- Estimate the materials released to the environment as a result of the failure or damage.
- Describe steps being taken to prevent recurrence.

In the event of a spill of drugs, pesticides or feed that results in a discharge to waters of the United States, you must provide an oral report of the spill to your permitting authority within 24 hours of occurrence and a written report in 7 days. The report must include:

- The identity of the material.
- The quantity spilled.

Refer to Appendix M for example forms that may be used to submit information about failure or damage to the structure of containment systems and spills of drugs, pesticides, or feed to your permitting authority.

📖 *Regulation: 40 CFR 451.3*

Examples of Information to Include When Reporting Structural Failure

- *Cause of the structural failure – storm broke a hole in 2 nets; raceway screens clogged and caused overflow.*
- *Time that elapsed until the failure was repaired – 2 hours until the nets were repaired; 30 minutes until the screen was unclogged.*
- *Amount and composition of the spill – 2 tons of feed were washed overboard in heavy seas; 1,200 1.8 pound steelhead escaped; or 150 pounds of medicated feed containing Terramycin (0.55% oxytetracycline) spilled into a raceway and discharged.*
- *Steps being taken to prevent recurrence: routinely inspect and perform maintenance on nets; clean screens regularly to prevent clogging.*

Chapter 7: Narrative Requirements for Flow-through, Recirculating, and Net Pen Facilities

EPA established narrative requirements for flow-through, recirculating, and net pen CAAP facilities. Under the requirements for flow-through and recirculating facilities, you must develop and maintain a BMP plan on site that describes how you will achieve the following requirements:

- Solids control
- Material storage
- Structural maintenance
- Record-keeping
- Training

The plan must be made available to your permitting authority upon request. You must also certify that a BMP plan has been developed. Guidance on developing and certifying a BMP plan is available in Chapter 8 of this document. An example form for certifying your BMP plan is available in Appendix F.

Under the narrative requirements for net pen facilities, you must develop and maintain a BMP plan on site that describes how you will achieve the following requirements:

- Feed management
- Waste collection and disposal
- Transport or harvest discharge
- Carcass removal
- Material storage
- Maintenance
- Record-keeping
- Training

The plan must be made available to your permitting authority upon request. You must also certify that a BMP plan has been developed. Guidance on developing and certifying a BMP plan is available in Chapter 8 of this guidance. An example form for certifying your BMP plan is available in Appendix F.

What is required for solids control?

The following is required for solids control for flow-through and recirculating facilities:

- Use efficient feed management and feeding strategies that limit feed input to the minimum amount reasonably necessary to achieve production goals and sustain targeted rates of aquatic animal growth.
- Identify and implement procedures for routine cleaning of rearing units and offline settling basins.
- Identify procedures for inventorying, grading, and harvesting aquatic animals that minimize discharge of accumulated solids.
- Remove and dispose of aquatic animal mortalities properly on a regular basis to prevent discharge to waters of the United States, except where authorized by your permitting authority in order to benefit the aquatic environment.

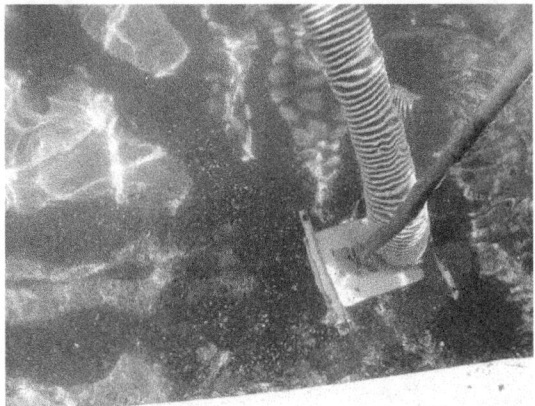

Figure 7.1. Cleaning out a raceway

Specific guidance for achieving solids control is available from Chapter 9 of this document. An example log for tracking and calculating feed conversion ratios is available in Appendix N.

📖 *Regulation: 40 CFR 451.11(a)*

What is required for materials storage?

The following is required for materials storage for flow-through, recirculating, and

net pen facilities:

- Ensure proper storage of drugs, pesticides, and feed in a manner designed to prevent spills that may result in the discharge to waters of the United States.
- Implement procedures for properly containing, cleaning, and disposing of any spilled materials.

Specific guidance for achieving materials storage is available from Chapter 10 of this guidance. An example log for tracking spills and leaks is available in Appendix O.

📖 *Regulation: 40 CFR 451.11(b) and 451.21(e)*

Figure 7.2. Storage container for an INAD drug

What is required for maintenance?

The following is required for structural maintenance for flow-through and recirculating facilities: FT RAS

- Routinely inspect production systems and wastewater treatment systems to identify and promptly repair damage.
- Regularly conduct maintenance of production systems and wastewater treatment systems to ensure their proper function.

The following is required for maintenance for net pen facilities: NET

- Routinely inspect production systems to identify and promptly repair damage.

- Regularly conduct maintenance of production systems to ensure their proper function.

Specific guidance for achieving maintenance is available from Chapter 11 of this document. Example inspection and maintenance logs for performing maintenance are available in Appendix P.

📖 *Regulation: 40 CFR 451.11(c) and 451.21(f)*

What is required for record-keeping?

The following is required for record-keeping:

- Maintain records for aquatic animal rearing units documenting feed amounts and estimates of the numbers and weights of aquatic animals in order to calculate representative feed conversion ratios.

- Keep records documenting frequency of cleaning, inspections, maintenance, and repairs.

- Keep records documenting net pen changes, inspections, and repairs. [NET]

Specific guidance for record-keeping is available from Chapter 12 of this guidance. A checklist of the record-keeping requirements of the CAAP ELGs is available in Appendix R. An example log for tracking cleaning of production systems and/or wastewater treatment systems is available in Appendix Q.

📖 *Regulation: 40 CFR 451.11(d) and 451.21(g)*

What is required for training?

The following is required for training:

- Train all relevant personnel in spill prevention and how to respond in the event of a spill to ensure proper clean-up and disposal of spilled materials. [FT] [RAS] [NET]

- Train personnel on proper operation and cleaning of production and wastewater treatment systems, including feeding procedures and proper use of equipment.

- Train personnel on proper operation and cleaning of production systems, including feeding procedures and equipment. [NET]

Specific guidance for training is available from Chapter 13 of this document. An example log for tracking training of employees is available in Appendix S.

📖 *Regulation: 40 CFR 451.11(e) and 451.21(h)*

What is required for feed management?

The following is required for feed management for net pen facilities: [NET]

- Employ efficient feed management and feeding strategies that limit feed input to the minimum amount reasonably necessary to achieve production goals and sustain targeted rates of aquatic animal growth.
- Minimize accumulation of uneaten feed beneath the pens through active feed monitoring and management

strategies approved by your permitting authority.

Specific guidance for feed management is available from Chapter 14 of this guidance. An example log for tracking and calculating feed conversion ratios is available in Appendix N.

🕮 *Regulation: 40 CFR 451.21(a)*

What are additional requirements for net pens?

The following requirements apply to net pen facilities: NET

- Collect, return to shore, and properly dispose of all feed bags, packaging materials, waste rope, and netting.
- Minimize any discharge associated with the transport or harvesting of aquatic animals (including blood, viscera, aquatic animal carcasses, or transport water containing blood).
- Remove and dispose of aquatic animal mortalities properly on a regular basis to prevent their discharge into waters of the United States.

Specific guidance related to the above net pen requirements is available from Chapter 15 of this document. An example log for tracking carcass removal and disposal is available in Appendix T.

🕮 *Regulation: 40 CFR 451.21(b), (c), (d)*

Chapter 8: Writing and Certifying a BMP Plan

A BMP is a means of controlling and reducing pollutant discharges other than those that rely on mechanic/physical or chemical systems. BMPs include schedules of activities, prohibition of action, maintenance and management procedures, and other treatment and operating requirements. These practices may be in addition to or separate from physical treatment systems.

Based on the CAAP regulation, facility operators must design BMP plans to include a series of practices such as feed management, feed monitoring, solids control, and material storage. BMP plans are flexible and allow facility operators to design measures and practices that work within their facility management framework.

> *Publicly supported aquaculture extension specialists, fact sheets, and reports are available to assist with developing BMP plans. These resources can help assure that the production practices, design parameters, and equipment will achieve the environmental protection goals being addressed.*

In the context of the CAAP ELGs, the BMP plan must include components that are designed to minimize the discharge of solids from the facility. The goal of the plan is to control conventional and nutrient pollutants in the discharge.

The CAAP facility is expected to provide written documentation of a BMP plan and keep necessary records to demonstrate the implementation of the plan. This type of regulatory structure allows individual facilities to develop a plan tailored to the unique conditions of the CAAP facility, while reducing the discharge of pollutants consistent with the goals of the CWA.

What is the goal of a BMP plan?

The goal of a BMP plan, as stated under the narrative requirements of the ELGs, is to describe the standard operating procedures and BMPs used to control solids, store materials, maintain the aquatic animal containment structures, perform record-keeping, train employees, monitor feeding, collect and dispose of waste, address transport or harvest discharge, and remove dead aquatic animals.

How do I write a BMP plan?

Your plan should describe how you will comply with each required element of the regulations. It should incorporate the following components, depending on the type(s) of system at your facility:

Solids Control

For your flow-through and/or recirculating systems, describe in detail how you will:

- Perform efficient feed management to limit feed input to the minimum amount reasonably necessary to achieve production goals and sustain targeted rates of aquatic animal growth.

- Identify and implement procedures for routine cleaning of rearing units and offline settling basins.

> *As part of your BMP plan, you should define the term "routine," which can vary during the year.*

- Identify procedures for inventorying, grading, and harvesting aquatic animals that minimize discharge of accumulated solids.
- Remove and dispose of aquatic animal mortalities properly on a regular basis to prevent discharge to waters of the United States (except where authorized by your permitting authority in order to benefit the aquatic environment).

Regulation: 40 CFR 451.11(a)

Material Storage

For your flow-through, recirculating and/or net pen systems, describe in detail how you will: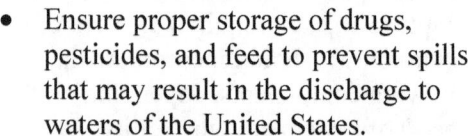

- Ensure proper storage of drugs, pesticides, and feed to prevent spills that may result in the discharge to waters of the United States.
- Implement procedures for properly containing, cleaning, and disposing of any spilled materials.

Regulation: 40 CFR 451.11(b) and 451.21(e)

Refer to the EPA Office of Pesticides website on Pesticide Storage Resources (http://www.epa.gov/pesticides/regulating/storage_resources.htm) or JSA's *Guide to Drug, Vaccine, and Pesticide Use in Aquaculture* at (http://aquanic.org/jsa/wgqaap/drugguide/drugguide.htm) for useful information or

suggestions for ensuring proper storage or drugs, pesticides, and feed to prevent spills.

Maintenance

For your flow-through and/or recirculating systems, describe in detail how you will:

- Routinely inspect production systems and wastewater treatment systems to identify and promptly repair damage.
- Regularly conduct maintenance of production systems and wastewater treatment systems to ensure their proper function.

For your net pen systems, describe in detail how you will:

> *As part of your BMP plan, you should define the terms "routinely" and "regularly," which can vary during the year.*

- Routinely inspect production systems to identify and promptly repair damage.
- Regularly conduct maintenance of production systems to ensure their proper function.

Regulation: 40 CFR 451.11(c) and 451.21(f)

Record-keeping

For your flow-through and/or recirculating systems, describe in detail how you will:

- Maintain records for aquatic animal rearing units documenting feed amounts and estimates of the numbers and weights of aquatic animals in order to calculate representative feed conversion ratios.

- Keep records documenting frequency of cleaning, inspections, maintenance, and repairs.

For your net pen systems, describe in detail how you will:

- Maintain records for aquatic animal rearing units documenting feed amounts and estimates of the numbers and weights of aquatic animals in order to calculate representative feed conversion ratios.
- Keep records documenting net pen changes, inspections, and repairs.

📖 *Regulation: 40 CFR 451.11(d) and 451.21 (g)*

Training

For your flow-through and/or recirculating systems, describe in detail how you will:
`FT` `RAS`

- Train all relevant personnel in spill prevention and how to respond in the event of a spill to ensure proper clean-up and disposal of spilled materials.
- Train personnel on proper operation and cleaning of production and wastewater treatment systems, including feeding procedures and proper use of equipment.

For your net pen systems, describe in detail how you will: `NET`

- Train all relevant personnel in spill prevention and how to respond in the event of a spill to ensure proper clean-up and disposal of spilled materials.

- Train personnel on proper operation and cleaning of production systems, including feeding procedures and equipment.

📖 *Regulation: 40 CFR 451.11(e) and 451.21(h)*

Feed Management

For your net pen systems, describe in detail how you will: `NET`

- Employ efficient feed management and feeding strategies that limit feed input to the minimum amount reasonably necessary to achieve production goals and sustain targeted rates of aquatic animal growth.
- Minimize accumulation of uneaten feed beneath the pens through active feed monitoring and management strategies approved by your permitting authority.

Documenting efficient feed management for EPA can be accomplished by describing the following:

- Feed methods used to minimize solids production.
- Modifications made to feed quantities as fish production changes (e.g., size, health of fish).
- Feed handling methods used to reduce generation of fines.
- Feed formulations information for each life-history stage of fish reared.

📖 *Regulation: 40 CFR 451.21(a)*

Waste Collection and Disposal

For your net pen systems, describe in detail how you will: `NET`

- Collect, return to shore, and properly dispose of all feed bags, packaging materials, waste rope, and netting.

📖 *Regulation: 40 CFR 451.21(b)*

Transport or Harvest Discharge

For your net pen systems, describe in detail how you will: [NET]

- Minimize any discharge associated with the transport or harvesting of aquatic animals (including blood, viscera, aquatic animal carcasses, or transport water containing blood).

📖 *Regulation: 40 CFR 451.21(c)*

Carcass Removal

For your net pen systems, describe in detail how you will: [NET]

- Remove and dispose of aquatic animal mortalities properly on a regular basis to prevent their discharge into waters of the United States.

📖 *Regulation: 40 CFR 451.21(d)*

Other Information

Including a diagram or map of the facility to illustrate the layout of the operation may be helpful to your permitting authority. Also include a statement certifying that the facility manager and the individuals responsible for implementing the BMP plan have reviewed and endorsed the plan.

A template for developing your BMP plan is available in Appendix E. A sample BMP plan and a checklist of components to include in your BMP plan are also available in Appendix E.

How do I certify my BMP plan?

Send a signed letter to your permitting authority, stating that you have developed a BMP plan. You will need to send a letter every time your permit is renewed. The BMP certification form should include your name and title, name of the facility, NPDES number, and date the BMP plan was developed. An example certification form that may be submitted to your permitting authority is available in Appendix F.

If you have any questions about certifying your BMP plan, be sure to check with your permitting authority.

Chapter 9: Solids Control for Flow-through and Recirculating Facilities

The ELGs regulations require facilities to implement solids control BMPs, which are intended to allow facilities to develop regional or site-specific operational measures to control the discharge of solids and other materials. The narrative BMP requirements also allow facilities and permit writers to respond to state programs that are currently working well. Some examples of solids control management practices include a combination of any and/or all of the following:

- Feed management
- Solids management
- Solids disposal
- Solids storage
- Mortality removal and disposal

Feed Management

Feed is effectively the only major source of aquaculture-derived nutrients, such as nitrogen and phosphorus, and solids in flow-through systems. Optimizing feed management by using high quality feeds and minimizing feed waste can reduce the nutrients and solids generated and released to the environment. Feed also represents the largest single variable cost of production and efficient use of feeds can result in cost savings. Accurate feeding systems and appropriate feeding levels are essential for productivity, economic efficiency, and protection of the environment.

Relatively short hydraulic residence times and continuous discharge of water make feed management an important component in controlling the amount of nutrients and

solids discharged from flow-through systems.

For recirculating aquaculture systems, the loading of potential pollutants to a receiving body of water is not entirely related to feed input, but is dependent upon the effectiveness of waste capture and treatment processes within the recirculating system and on any additional effluent treatment processes used to clean the water before discharge. Minimizing waste feed will minimize the wastes that must be treated in the recirculating system and ultimately the amount of waste released to the environment. Feed management is only one factor among many in the control of potential pollution from recirculating aquaculture systems.

Examples of Feed Management Practices

1. Use high quality feeds and seek to minimize nutrient and solids discharges through optimization of feed formulation (in cooperation with feed manufacturers)

Feeds should be formulated to meet the nutritional requirements of the cultured species and for optimum feed conversion ratios and retention of protein (nitrogen) and phosphorus. Feeds should be formulated using ingredients that have high dry matter and protein apparent digestibility coefficients. Formulations should be designed to enhance nitrogen and phosphorus retention efficiency, and reduce metabolic waste output. Feeds should contain sufficient dietary energy to spare dietary protein (amino acids) for tissue synthesis. Available phosphorus levels should be slightly in excess of the dietary requirements of the species for each life-history stage. Efforts should be made in feed formulation to keep total phosphorus levels as low as possible while maintaining appropriate available phosphorus levels. Consult a qualified aquatic animal nutritionist or feed manufacturer for information regarding feed formulation.

When facility operators evaluate feed formulations, they should consider numerous factors including, pellet stability, digestibility, palatability, sinking rates, energy levels, moisture content, ingredient quality and the nutritional requirements of the species being grown. Feeds should be formulated and manufactured using high-quality ingredients.

Pelleted feeds should be stable in water for sufficient time so the pellets remain intact until eaten. Feeds should be manufactured, stored, shipped, and handled at the farm so they contain a minimum amount of fine particles.

In the case of feeds used in recirculating systems, minimizing metabolic excretion of nitrogen from amino acids catabolized to provide metabolic energy, and minimizing nitrogen excretion in feces from indigestible protein is the top priority in feed formulation. Therefore high quality feeds for recirculating systems should have balanced amino acid profiles, e.g., profiles that meet but do not substantially exceed dietary requirements for individual essential amino acids, and contain sufficient dietary energy from carbohydrates and lipids to "spare" dietary protein for tissue synthesis.

2) Use efficient feeding practices

In flow-through and recirculating systems, feed can be delivered by hand, automatic feeders, demand feeders, or by mechanical feeders. Regardless of the delivery method or system, the amount of feed offered should optimize balance between growth goals and feed efficiency.

Figure 9.1. Demand feeder

The appropriate quantity of feed for a given species is influenced by feed formulation, fish size, water temperature, dissolved oxygen levels, carbon dioxide concentrations, health status, and management goals. Feed particle size should be appropriate for the size of fish in each rearing unit. Whenever possible, feed utilization should be monitored by observing feeding behavior or by looking for trends in waste feed collecting within the culture unit or waste feed exiting the culture unit.

Figure 9.2. Feed truck

Multiple feeding periods distributed over a 24-hour period will provide more uniform water quality within a recirculating system than a feeding schedule only offering feed once or twice daily.

3) Calculate feed conversion ratios by using feed and fish biomass inventory tracking systems FT RAS ✓

Calculation of feed conversion ratios is an essential function on all aquatic animal farms. Monitoring long- and short-term changes in feed conversion ratios allows farmers to quickly identify significant changes in feed consumption and waste production rates.

4) Manage within the carrying capacity of the production system FT RAS ✓

Flowing water carries dissolved oxygen to the culture units, receives the waste produced in the culture unit, and carries these wastes away from the culture unit to treatment units before the wastes can accumulate to harmful and undesirable levels. Dissolved oxygen is usually the first water quality parameter to limit culture tank carrying capacity, which, in simplistic terms, is the maximum fish biomass that can be supported at a selected feeding rate. Note that the culture vessel volume does not determine carrying capacity unless water flow is in excess of all other water quality based carrying capacity requirements.

Impaired water quality due to high loading and excessively high feeding rates stresses fish and reduces feed efficiency and production. High loading and high feeding rates also lead to higher levels of nutrients and solids in the effluent. Loading and

feeding rates within the carrying capacity of the production system are more efficient, and minimize the discharge of pollutants.

In flow-through systems, carrying capacity is determined by incoming water quality and quantity, production goals, facility design, site characteristics, and species cultured. As such, there is no single carrying capacity value applicable to flow-through systems and carrying capacity will vary within and between facilities.

Recirculating systems, by definition, treat and reuse large portions of the system make-up water flow. Therefore, the water flow requirements through the culture units within a recirculating system can be much greater than the make-up water flow requirements that flush the system. Of primary importance is the removal of the waste metabolites: ammonia, carbon dioxide, and total suspended solids (TSS), whose production is directly proportional to feed load.

Biofilters, aeration columns, and filters/clarifiers are unit processes used to control ammonia, carbon dioxide and TSS accumulations within recirculating systems. Aquacultural engineering texts and many other publications provide the methodology to design biofilters, aeration columns, and filters/clarifiers to treat a given flow or the waste metabolites produced by a given feeding rate. However, when a unit treatment process (e.g., biofilter or aeration column) is designed, the designer should provide the expected water quality exiting a culture tank within a recirculating system to help ensure that the design will provide safe water quality for the fish when reared at maximum carrying capacity, i.e., feed loading. Operators of recirculating systems should feed at rates that do not exceed the maximum carrying capacity of the system.

5) Properly store feed FT RAS

Feed storage areas should be secure from contamination, vermin, moisture and excessive heat. Long term storage of feed can affect feed quality. Feed should be rotated (use oldest feed first) and not stored beyond the manufacturer's recommended use date. If feed can no longer be used because of spoilage or it has exceeded the manufacturer's recommendations for storage, the unusable feed should be properly disposed to prevent water quality impacts.

Care should be taken during feed handling to minimize pellet damage or crushing and reduce the creation of fine feed particles that cannot be utilized by the fish.

6) Check feeding equipment to ensure efficient operation FT RAS

Improperly adjusted or malfunctioning feeding equipment can over-feed or under-feed fish and reduce feed and production efficiency.

7) Conduct employee training in fish husbandry and feeding methods to ensure that workers have adequate training to optimize feed conversion ratios FT RAS

Solids Management

Fish fecal matter and waste feed are the major constituents of total suspended solids from culture practices in flow-through systems. Solids allowed to settle in rearing units degrade water quality and may irritate fish gills leading to disease. Solids can impact the aquatic environment and should

be thoroughly collected prior to wastewater discharge.

Flow-through system effluent is characterized as high volume with low solids concentration. This effluent characteristic generally limits practical and economical solids management to the capture and removal of solids using settling basins. Solids found in flow-through systems readily settle and can be managed with practices that rely on gravitational settling before water is discharged. Practices that increase solid particle fragmentation decrease settling efficiency. These particles are much smaller and have poor settling characteristics. Fish grading, harvesting, and other activities within raceways or ponds should be conducted in a manner that minimizes the disturbance and possible discharge of accumulated solids. In rare situations, high levels of TSS in source water may warrant pretreatment systems, such as settling basins, to improve source water quality for fish culture.

Waste feed and fish fecal matter are waterborne and require separation for efficient management of water quality within the recirculating system. The solids treatment processes in a recirculating system remove a portion of the feed derived waste solids in the recirculating water. Higher solids removal efficiencies result in cleaner water within the recirculating system.

The particulate wastes discharged from the recirculating system are contained in either a small but concentrated flow (such as the intermittent backwash from a solids capture unit) and/or in a more continuous flow of displaced water (such as an overtopping flow from a pump sump that is water that has been displaced by make-up water addition) that has a concentration of solids similar to that found in the fish culture tanks. Not all recirculating systems will have an overtopping flow, depending upon their make-up water requirements. When the solids are discharged, as with backwashing water, the concentration of solids is typically relatively high. However, if an overtopping flow is discharged from the system, it will be relatively small in volume compared to the discharge from a flow-through system.

Solids can impact the aquatic environment and should be collected as much as possible prior to wastewater discharge. Therefore, nearly all flow-through and recirculating aquaculture systems use some form of solids treatment and/or disposal to remove the concentrated slurry of captured biosolids. In some cases, it may also be necessary to treat the more dilute but relatively larger volume system overflow before this flow is discharged. As an alternate to on site treatment, either of these waste flows could be discharged to a POTW.

Examples of Solids Management Practices

1) Design and operate rearing units for efficient and rapid capture of solids from the water column, incorporating fish-free settling basins where practical FT RAS

Linear rearing units designed to promote plug flow and sufficient water velocity to prevent the settling of solids within the rearing unit allow the efficient capture of solids using quiescent zones or other settling basins. Proper facility design and construction can be an economical means of managing solids through settling in designated areas, allowing for efficient removal. Fish should be prevented from entering quiescent zones and other settling basins and removed as soon as possible when found to prevent or alleviate

resuspension and subsequent discharge of solids.

Circular tanks with properly designed inlets and drains can remove the majority of solids with minimum labor for further treatment. Circular tanks can rapidly concentrate and remove settleable solids. Circular tanks are designed to promote a primary rotating flow that creates a secondary radial flow that carries settleable solids to the bottom center of the tank, making the tank self-cleaning. The self-cleaning attribute of the circular tank depends on the overall rate of flow leaving the bottom-center drain, the strength of the bottom radial flow towards the center drain, and the swimming motion of fish. The factors that affect self-cleaning within circular tanks are also influenced by the water inlet and outlet design, tank diameter-to-depth ratio, water rotational period, size and density of fish, size and specific gravity of fish feed and fecal material, and water exchange rate.

2) Remove solids from collection systems in a timely fashion FT RAS

Solids should be removed from quiescent zones with a frequency sufficient to prevent cohesion and limit release of solids-bound nutrients. The level of feed application, settling basin efficiency, and relative storage capacities of the basins will determine the removal frequency. For example, quiescent zones are typically cleaned at least every two weeks during the peak growing seasons. However, the frequency of solids removal should be determined based on factors such as facility discharge compliance limits, water quality requirements in the culture units, or labor availability.

The most common method of solids removal from quiescent zones is by suction through a vacuum head. Usually, a standpipe in each quiescent zone connects to a common pipe that carries the slurry to an off-line settling basin. Suction is provided by head pressure from raceway water depth and gravity, or where fall is not available, by pumps. A flexible hose and swivel joint connects the vacuum head to the standpipe so the vacuum can be manipulated to clean the quiescent zone. There are other methods used to clean quiescent zones. For example, the standpipe to the off-line destination may be removed and the solids can be pushed with a broom or squeegee device to the suction port.

Settling basins should be cleaned as frequently as practical. A procedure or mechanism to remove the dewatered manure from the thickening device must be incorporated. Sludge left too long in settling basins becomes sticky and viscous making removal more difficult. Sludge accumulation may degrade water quality as nutrients are released through bacterial or physical degradation. Off-line and full-flow settling basins should be harvested when storage capacity is reached or as effluent concentrations near compliance limits.

To clean full-flow and off-line settling basins the inflow is usually diverted to another settling basin and the supernatant from the settling basin decanted. The slurry is allowed to dry sufficiently for removal by backhoe, front-end loader, or other equipment. Or the slurry may be pumped directly onto a tank truck or manure spreader. Other options include pumping the slurry out of the settling basin without diverting the flow, similar to cleaning a quiescent zone.

3) Remove solids rapidly, but gently

Rapid, effective, and gentle removal of waste solids within a solids treatment unit is the best approach to use when targeting optimum water quality. Waste feed and fish manure are typically fragile and labile organic particles. The longer these particles are held within the culture system, the more opportunity that dissolution forces such as hydraulic shear and micro-organisms will have to disintegrate larger particles into much finer and more soluble particles. Finer particles can more rapidly leach nutrients and biochemical oxygen demand (BOD) and these components are harder to remove from the water column than the original intact fecal pellet or waste feed pellet. Thus, if unit processes are not installed to remove fresh and intact solids rapidly, then solids decomposition within aquaculture systems can degrade water quality and thus directly affect fish health and the performance of other unit processes. Products of solid decomposition are more difficult to remove from aquacultural effluents.

Waste solids exiting the rearing tank can be removed from the bulk flow leaving the culture tank using a treatment unit such as settling basins (e.g., full-flow settlers, off-line settlers, quiescent zones, inclined [tube or plate] settlers, and swirl separators), microscreen filters (e.g., drum, disk, or belt filters), and granular media filters (e.g., bead or sand filters). In addition, ozone and foam fractionation are water treatment processes that can be used in recirculating systems to remove dissolved organic matter.

Conventional sedimentation and microscreen filtration processes are often used to remove solids larger than 40-100 µm. However, few processes used in aquaculture can remove dissolved solids or fine solids smaller than 20-30 µm, although granular media filtration has been used to remove these fine solids. Depending on the particle size distribution and the concentration of solids, conventional sedimentation and microscreen filtration processes typically remove anywhere from 30-80% of the solids in the treated flow.

Significant degradation or re-suspension/flotation of the solids matter should be avoided, but can occur in treatment units that have relatively infrequent backwash cycles. Therefore, the best solids removal processes remove solids from the system as soon as possible and expose solids to the least turbulence, mechanical shear, or micro-biological degradation. Note that microscreen filters and swirl separators (with a continuous underflow) do not store solids for an appreciable period, unlike settling basins and most granular media filters.

Backwash of the solids capture unit will create an intermittent solids-laden flow that will require treatment before discharge, unless discharged to a POTW.

Not all recirculating aquaculture systems maintain low levels of suspended solids, as is typically the goal in recirculating systems used for sensitive species such as trout and salmon. Some species may tolerate elevated levels of suspended solids and may actually consume the algae or micro-organisms found in these solids. Such is the case for some recirculating systems used for tilapia and shrimp. Some recirculating systems rely on a combination "green water" or organic detrital algae soup (ODAS), which is an algal and activated sludge-type treatment process, combined with settling basins or granular media filters to treat the water. In these instances, the rapid removal of waste solids is not a goal because the algae and bacteria growing *in situ* within the recirculating systems may rely upon solids degradation to treat dissolved wastes and maintain the culture system water quality. Total suspended solids concentrations in

these recirculating systems can exceed 150 mg/L. Thus, the associated waste management systems must consider the specifics of each recirculating aquaculture system in order to successfully achieve waste collection, transfer, storage, treatment, and utilization.

4) Frequently remove solids from settling basins

Settling basins should be cleaned as frequently as practical. Fish manure left too long in settling basins becomes sticky and viscous making removal difficult. Fish manure accumulation may degrade water quality and can provide a substrate for bacterial growth.

A procedure or mechanism to remove the dewatered manure from the settling basin should be incorporated in the BMP plan.

5) Overflows from solids thickening tanks may require additional treatment
FT RAS

Solids thickening and storage tanks will often discharge a supernatant/overflow, which will be a relatively small volume discharge but one that contains the highest concentration of wastes discharged from an aquaculture system. Therefore, treating the thickening tank overflow can reduce the mass load of wastes discharged. Treatment can be relatively simple and inexpensive because low effluent volumes must be treated. Further removal of soluble BOD and ammonia may be required, and can be accomplished with properly designed aerated basins, aerobic lagoons, created wetlands, anaerobic filters, or other suitable technologies. Alternatively, the thickening tank overflow could be discharged to a

POTW or reused beneficially for irrigation or hydroponics.

6) When necessary, remove solids from the recirculating system's overtopping flow (if present) before it is discharged

Depending upon their make-up water requirements, some recirculating systems will have an overtopping flow in addition to a concentrated backwash flow. The concentration of solids in the overtopping flow is typically similar to that found in the fish culture tanks. Depending upon the specifics of the recirculating aquaculture system, the suspended solids in the flow overtopping this system may require further treatment. Waste solids can be removed from the overtopping flow using a treatment unit such as settling basins (e.g., full-flow settlers, inclined [tube or plate] settlers, and swirl separators), microscreen filters (e.g., drum, disk, or belt filters), granular media filters (e.g., bead or sand filters), or dissolved air flotation systems.

Solids Disposal

Aquaculture solids, primarily consist of fish feces and uneaten feed, contain plant nutrients (nitrogen and phosphorus) and can often be used as a soil amendment. The composition of the solids varies among facilities according to feed formulation(s) used at the facility, treatment of the solids inside and out of the culture system, and age of the solids.

Aquaculture solids are similar to other animal manures. Some state or local governments may consider the fish manure captured in an aquaculture waste management system an industrial or municipal waste (i.e., not an agricultural waste). Check with your local or state

authorities to determine your waste disposal options.

Examples of Solids Disposal Practices

1) Disposal of solids should comply with all applicable local and state regulations and done in a manner that prevents the material from entering surface or groundwaters

Solids disposal will be a site-specific practice, based on factors such as local regulations, soil types, topography, land availability, climate, and crops grown. Disposal options might include land application on agricultural lands at agronomic rates, storage lagoons, composting, and contract hauling.

a) Land Application – Land application of aquacultural solids is the most common disposal method. Proper application of aquacultural solids provides a safe method for solids utilization while fertilizing crops and amending the soil. Fish manure in liquid form may be sprinkler irrigated directly onto agricultural land. In slurry form, fish manure may be pumped into a tank truck or manure spreader and then applied to agricultural land. Finished compost generated from aquacultural solids may also be applied onto agricultural land at agronomic rates.

Figure 9.3. Manure spreader

b) Evaporation Ponds/Lagoons – Manure slurries from aquaculture operations may be treated in evaporation ponds/lagoons that can thicken and stabilize the manure. Evaporation ponds/lagoons are effective in arid climates only. The stabilized solids can then be land applied or otherwise safely disposed.

c) Composting – Thickened and dewatered manure may be composted. Composting stabilizes the solids and produces a valuable soil amendment. Aerobic static pile composting is the most common method for composting dewatered manure. Any excess supernatant, leachate, or filtrate leftover from slurry treatment processes may require additional treatment. State and local regulations regarding composting should be followed.

d) Contract Hauling – A licensed contract hauler can be paid to remove the collected solids or thickened manure.

e) Reed Drying Beds – Depending on location and the local regulations, an aquaculture facility may have only limited and costly options available for disposal of the thickened manure. For example, transportation costs may make sludge disposal on cropland uneconomical. Disposing of the sludge on-site within created wetlands may be an attractive alternative.

A constructed reed drying bed can provide on-site treatment of a concentrated solids discharge with an uncomplicated, low-maintenance, plant-based system. Reed drying beds are vertical-flow wetland systems that have been used over the past 20 years to treat thickened sludge (1-7% solids) produced in the clarifier underflow at wastewater treatment plants and have been recently used to treat manure from commercial recirculating systems. Thickened biosolids are loaded in sequential batches onto the reed drying bed every 7-21 days. Only 2-4 inches of thickened biosolids are applied during a given application. The 1-3 week intervals between applications of thickened biosolids allow for dewatering and drying, which is facilitated by the vegetation growing on the sand bed. Reed beds have been found to have a useful lifetime of up to 10 years.

2) Use solids from earthen flow-through systems to repair embankments FT

Earthen flow-through systems accumulate solids in the rearing units during production. It is not practical or necessary to remove solids during production from earthen rearing units. When it is necessary to remove solids from earthen rearing units, the source water is diverted around or away from the rearing units and they are allowed to dry. The solids removed can be used to repair the embankments and other areas of the rearing unit. Solids should not be used to repair roads or other facility surfaces because the solids could contaminate stormwater runoff from the facility.

Solids Storage

Concentrated aquaculture solids can be stored in thickening basins that have been designed to accommodate the build-up of solids and hence provide some temporary solids storage capacity. However, solid-liquid separation becomes less effective as sludge accumulates within these basins. Increasing sludge depths can compromise settling basin hydraulics and the solids stored can rapidly ferment leading to solids flotation and dissolution of nutrients and organic matter. In some cases, the thickened sludge from thickening basins is transferred to large sludge storage structures capable of holding months of captured and thickened solids. Facility owners design solids storage structures to hold the amount of solids that they anticipate they will collect and hold at the facility prior to final disposal (e.g., land application). Facilities located in colder climates may be required to hold solids for 6 months, while facilities in warmer climates typically design storage structures to hold solids for one month. These off-line storage structures typically have zero overtopping flow and store their manure slurry contents until they can be removed for disposal.

Figure 9.4. Above ground manure storage structure

Examples of Solids Storage Practices

1) Store sludge in an appropriate facility or container FT RAS

Sludge storage structures include earthen ponds, and aboveground or belowground tanks. Earthen ponds are generally rectangular basins with inside slopes (horizontal:vertical) of 1.5:1 to 3:1. Depending on site geology and hydrology, earthen ponds can have liners of concrete, geomembrane, or clay. Because they are uncovered, earthen pond design will include the capacity for storage of rain water as well as a method for removing solids. In the case where solids will be removed via pumping, the solids must be agitated to provide a uniform consistency. Pond agitation may be accomplished with hitch-type propeller agitators that are powered by tractors or by agitation pumps. Propeller agitators work well for large ponds, while chopper-agitator pumps work well for smaller ponds. Solids removal may also be done with heavy equipment, in which case, pond design should include ramp access (maximum slope of 8:1) and suitable load capacity in the unloading work area.

Sludge may also be stored in tank structures, above and below ground. Storage tanks are primarily constructed of reinforced concrete, metal, and wood. Reinforced concrete tanks may be cast-in-place, walls, foundation, and floor slab, or they may be constructed of pre-cast wall panels, bolted together, and set on a cast-in-place foundation and floor slab. Metal tanks are also widely used, with the majority being constructed of glass-fused steel panels that are bolted together. There are many manufactured, modular tanks commercially available in reinforced concrete and metal, as well as wood.

Solids degradation during storage can produce dangerous levels of hydrogen sulfide gas, methane and hydrogen gases, and in tanks with little air exchange can contain an atmosphere that includes the aforementioned gases and is anoxic. Use OSHA confined space guidelines when considering all aspects of the human interface with a solids storage structure and take every practical precaution to prevent harm to those working around these structures.

State and local regulations regarding odors from the manure storage vessels should be considered.

Mortality Removal and Disposal

Mortality of small numbers of cultured species in aquaculture systems is a common occurrence. It is also unpredictable and highly variable among rearing units, epizootics, and facilities. A facility may experience chronic mortality of a few fish per day or a catastrophic loss caused by infectious disease or acute environmental stress. Depending on water temperature and species, dead fish either float or sink after dying, with warm water fish typically floating and cold-water fish sinking.

In flow-through systems, whether floating or sinking, dead fish tend to accumulate on the screens at the end of the rearing units.

In recirculating systems, sinking fish mortalities tend to accumulate on the exclusion screen on the bottom center drain of circular tanks or on the outlet screen of linear raceways. Floating fish will accumulate on the surface of circular tanks, where they are relatively easy to see.

The timely removal of mortalities helps decrease the probability of spreading

infectious organisms and the introduction of excess nutrients into the system.

Examples of Mortality Removal and Disposal Practices

1) Remove mortalities from rearing units on a regular basis

Mortalities should be removed from rearing units regularly. To accomplish this, inspect culture units to check for the presence of mortalities. Many mortalities float to the surface of the culture water and can be collected by hand or with nets.

Mortalities accumulating on screens prevent the efficient flow of water from unit to unit and represent a hazard for possible damage to the screen resulting in escape of fish from the unit or diversion of flow away from downstream units.

Dead or moribund fish can be transported by flowing water to a tank drain, where they can accumulate against screens and restrict the water flow out of the culture unit. Dead fish should be removed from culture units as soon as possible to maintain water level in the culture tank, to decrease the probability of spreading infectious organisms, and to reduce water quality deterioration. Dead fish that sink may be difficult to detect at the bottom center of large circular culture tanks or along the bottoms of net pens that are deep or contain turbid water. A procedure or mechanisms should be identified for detecting and removing dead fish from the culture units.

2) Follow recommended aquatic animal health management practices

Prevention and minimization of mortalities through proper fish health surveillance and management are the best methods for managing mortalities. Maintaining good water quality can help to prevent disease outbreaks. Most states offer diagnostic services and treatment recommendations for disease problems.

3) Do not discharge mortalities into receiving waters FT RAS NET

Appropriate screens for flow-through and recirculating systems on the outlet to receiving waters will prevent mortalities discharging into receiving waters. There are, however, permitted restoration and stock-enhancement activities where spawned carcasses are returned to waters for nutrient replacement.

Figure 9.5. Screened effluent pipe

4) Only use approved methods of mortality disposal FT RAS NET

Disposal methods are site-specific and usually governed by state or local regulations. Disposal options could include composting, rendering, use as fertilizer, incineration, burial, or landfill.

Chapter 10: Material Storage for Flow-through, Recirculating, and Net Pen Facilities

Material Storage

It is important to properly store materials used at aquaculture facilities to protect the environment. Specifically, the ELGs require that facilities ensure proper storage of drug, pesticides, and feed in a manner designed to prevent spills that may result in a discharge of these materials to waters of the United States. The ELGs also require facilities to implement procedures for properly containing, cleaning, and disposing of any spilled materials.

Examples of Material Storage Practices

1) Use and store drugs and pesticides in a manner to prevent contamination of the environment

Drugs and pesticides should be stored away from rearing areas, feeds, and water sources, in locations that are secure, dry, void of drains, water tight, well-ventilated, and not subject to extreme temperatures. Also consider securing the storage areas to avoid tampering or vandalism.

Refer to EPA's Office of Pesticides website on Pesticide Storage Resources (http://www.epa.gov/pesticides/regulating/storage_resources.htm) or JSA's *Guide to Drug, Vaccine, and Pesticide Use in Aquaculture* (http://aquanic.org/jsa/wgqaap/drugguide/drugguide.htm) for useful information or suggestions for ensuring proper storage or drugs, pesticides, and feed to prevent spills.

> *Drugs and pesticides should be used only when needed and only for the specific use indicated on the label. In some cases, drugs are used under an INAD exemption or prescribed by a veterinarian as an extralabel drug use. Use of these materials is regulated by federal and state agencies, and individuals are responsible for using these products according to label directions and disposing of containers and unused chemicals according to applicable federal and state regulations.*

2) Use and store feed in a manner to prevent contamination of the environment and to protect the quality of the feed

Feed should be stored away from rearing areas and water sources, in locations that are secure, dry, water tight, and not subject to extreme temperatures.

Storing feed properly maintains feed quality. To protect feed quality, store it to prevent insect and rodent contamination. Bacteria and fungi (mold) can destroy the nutritional value of feed and produce toxins, which may stress or kill fish. Keeping feed dry and maintaining temperatures to prevent condensation helps to minimize the growth of bacteria and molds. Follow the feed manufacturer's storage recommendations for best results.

Handle and store feed with care to prevent physical breakdown of feed into fine particles. If fines are present in feed, they should be removed and disposed of properly.

Although most currently used formulations are extruded pellets, which produce very little fines, check with your feed manufacturer to determine if they will provide a credit and take back the fines.

Figure 10.1. Feed storage area

3) Develop a spill response and prevention plan for drugs, pesticides, and feed (you can also develop these plans for petroleum products and other hazardous products that may be found at your facility)

The best way to avoid runoff contamination from spilled materials is to prevent the spill from occurring. Carefully storing materials in sound, clearly labeled containers and regular inspection and maintenance of equipment are key practices to prevent spills. Materials stored outdoors should be covered and kept on paved areas to protect them from being mobilized by wind and runoff. If not covered, storage areas should be designed to drain with a slight slope (approximately 1.5 percent) to an area that will provide treatment prior to disposal. Use secondary containment, such as berms, safety storage cabinets, or drum containment systems, when storing liquids.

State and federal laws require reporting of significant spills of many chemical products. Although the quantity of drugs and pesticides used and stored at CAAP facilities is generally small, check with state and local authorities for specific details about any chemicals that would require reporting in the event of a spill at your facility. A plan should be developed specifying response procedures, key staff, and phone numbers of regulatory authorities. All facility employees should be aware of the plan and the plan should be accessible to all employees at all times. Refer to Chapter 13 of this guidance for information about training employees in spill prevention.

Spill response and prevention plans can be used to ensure that a facility properly contains, cleans, and disposes of spilled materials. The plan should clearly state measures to stop the source of a spill, contain the spill, clean up the spill, dispose of contaminated materials, and train personnel to prevent and control future spills.

To develop the plan, first identify potential spill or source areas, such as loading and unloading, storage, and processing areas, and areas designated for waste disposal.

Provide documentation of spill response equipment and procedures to be used, ensuring that procedures are clear and concise. Give step-by-step instructions for the response to spills at a particular facility. This spill response and prevention plan can be presented as a procedural handbook or a sign. The spill response and prevention plan should:

- Identify individuals responsible for implementing the plan.
- Define safety measures to be taken with each kind of waste.

- Emphasize that spills must be cleaned up promptly.
- Specify how to notify appropriate authorities, such as police and fire departments, hospitals, or publicly-owned treatment works for assistance.
- State procedures for containing, diverting, isolating, and cleaning up the spill.
- Describe spill response equipment to be used, including safety and cleanup equipment.

The use of water for cleanup should be strongly discouraged. Launderable or disposable shop rags should be used for small spills of non-volatile chemicals, and rags should be properly cleaned or disposed of. Larger spills should be absorbed with vermiculite, sawdust, kitty litter, or absorbent "snakes." Disposal methods depend on the hazard level of the spilled material. Nonvolatile liquids can be cleaned up with a wet/dry shop vacuum and disposed of with the rest of the facility's waste. Drains or inlets to storm sewers should be plugged during spill remediation to prevent off-site runoff/discharge of pollutants.

A spill prevention and response plan must be well planned and clearly defined so that the likelihood of accidental spills can be reduced and any spills that do occur can be dealt with quickly and effectively. Training might be necessary to ensure that all relevant personnel are knowledgeable enough to follow procedures. Equipment and materials for cleanup must be readily accessible and clearly marked for personnel to be able to follow procedures.

Remember to update the spill prevention and response plan to accommodate any changes in the site or procedures. It is also important to regularly inspect areas where spills might occur to ensure that procedures are posted and cleanup equipment is readily available.

A spill prevention and response plan can be highly effective at reducing the risk of surface and groundwater contamination. However, the plan's effectiveness is enhanced by worker training, availability of materials and equipment for cleanup, and extra time spent by management to ensure that procedures are followed.

Spill prevention and response plans are inexpensive to implement. However, extra time is needed to properly handle and dispose of spills, which results in increased labor costs.

If you want to track spills from your facility for your own record-keeping, you can use the example tracking worksheet in Appendix O.

Additional Suggestions

1) Use and store petroleum products to prevent contamination of the environment

State and federal laws require reporting of significant spills of petroleum products. A plan should be developed specifying response procedures, key staff, and phone numbers of regulatory authorities.

Petroleum leaking from storage tanks or farm equipment wastes a valuable resource and can contaminate surface or underground water supplies. Petroleum products are highly odorous and small amounts in water can produce an off-flavor in aquatic animals. Petroleum storage in above-ground and underground tanks is regulated by federal and state agencies. Information on petroleum storage regulations can be obtained from state Departments of Commerce, state Departments of Environmental Quality or Protection, or from EPA regional offices. Aquaculturists should also implement a regular maintenance schedule for tractors, trucks, and other equipment to prevent oil and fuel leaks. Used oil should be disposed of through recycling centers.

Figure 10.2. Fuel storage

Facilities can also address spill prevention and response for petroleum products in their spill prevention and response plan, described above.

Chapter 11: Maintenance for Flow-through, Recirculating, and Net Pen Facilities [FT] (RAS) [NET]

Maintenance

Flow-through, recirculating, and net pen systems should be well-maintained, managed efficiently, and operated in compliance with all applicable laws and regulations. This will improve long-term economic performance and reduce environmental impact. As such, the following management practices are simply part of good management.

Examples of Maintenance Practices

1) Maintain structures and equipment to ensure staff safety and protection of the environment [FT] (RAS) [NET]

Routinely inspect flow-through and recirculating production systems and wastewater treatment systems to identify and promptly perform repairs or replacement, as necessary.

Some of the system components that should be considered for routine inspection of flow-through and recirculating systems include:

- Drains—make sure that all of the parts of the drain structure are properly functioning; look for the proper placement of stand pipes, dam boards, and animal exclusion devices (for example screens across pipe openings); check that valves and other critical drain components are working properly; check for broken parts and repair when necessary.
- Production units—make sure that tanks and raceways are structurally sound; repair cracks as necessary; all

plumbing components are installed and working properly.
- Life support systems—routinely inspect oxygen equipment, filters, heaters, and any other life support equipment used to maintain optimal growing conditions.
- Feeding equipment—test automatic and mechanical feeders periodically to ensure that they are delivering the proper amounts of feed; check demand feeders for proper operation and adjust as necessary; inspect all feed storage areas to make sure that the feed is not contaminated by foreign substances, is not easily accessible to rodents and insects, and check for excess moisture and water leaks to prevent mold from forming.
- Solids control equipment and systems—check quiescent zones for proper function; inspect drains for clogging; and make sure that all settling basins are working properly and that the structures are safe and secure to prevent spills and accidental discharges of collected solids due to cracked or damaged basin structures.

Routinely inspect net pen systems to identify and promptly perform repairs or replacement of nets.

Some of the system components that should be considered for routine inspection at net pen facilities include:

- Nets—inspect for holes and physical damage to the nets and make sure that nets are securely attached to floating structures; if present, maintain predator control nets and devices to ensure proper operation.

- Floating structures—inspect for physical damage that may lead to structural failure during storms or periods of icing; check all mooring lines and anchor points for proper function and physical damage.
- Feeding equipment—test automatic and mechanical feeders periodically to ensure that they are delivering the proper amounts of feed; check demand feeders for proper operation and adjust as necessary; inspect all feed storage areas to make sure that the feed is not contaminated by foreign substances, is not easily accessible to rodents and insects, and check for excess moisture and water leaks to prevent mold from forming.

An example log for documenting routine inspections and repairs is available in Appendix P.

2) Periodically conduct a systematic review of your current facility to identify any problems that would lead to environmental impacts; when considering modifications to existing facility components or operations, include a review of the type and extent of probable environmental impacts that may occur as a result of the new methods

3) Clearly mark all net pen sites in accordance with the farm's permit for fixed private aids (buoys, navigation lights, etc.) to navigation from the U.S. Coast Guard and appropriate state authorities; make sure all net pen sites continue to be clearly marked in accordance with U.S. Coast Guard marking regulations NET

4) When installing net pens and their associated mooring systems, give careful consideration to their potential impacts on water circulation patterns; gear deployment should seek to optimize circulation patterns and maximize water exchange through the pens, thereby improving fish health and reducing benthic impacts NET

5) Design, operate, and maintain all holding, transportation, and culture systems to function as designed FT RAS

For flow-through systems, screens of appropriate size and strength should be installed at the intake from the source and outlet to receiving waters to prevent loss and escape of cultured species. Occasionally check screens to ensure that debris is not blocking them.

For recirculating systems, barriers of appropriate size and strength should be installed on the facility discharge and on the make-up water entry into the facility. A procedure or mechanisms should also be identified to prevent debris from plugging the barriers, thus preventing water from overflowing or bypassing the screens.

6) Avoid siting facilities in areas prone to frequent flooding FT RAS

Floods that overflow flow-through and recirculating systems result in loss of cultured animals and are usually catastrophic for the farmer. Facilities adjacent to surface waters should be constructed to minimize the possibility of flood waters entering the facility.

7) Transfer fish (stocking, grading, transfer, or harvest) in appropriate weather conditions and under constant visual supervision of at least one person; use appropriate equipment for the weather and cage designs; use shields or additional nets to prevent stray fish to escape during transfer (where necessary or appropriate) NET

8) Only obtain nets from a manufacturer or supplier whose equipment design specifications and manufacturing standards meet generally accepted standards prevalent in the aquaculture industry NET

Net design and specification should be commensurate with the prevailing conditions of the site. Stress tests should be preformed on all nets with more than three years of use in the marine environment when the net is pulled out and cleaned. All nets in use should be UV-protected.

9) Only obtain net pen structures from a manufacturer or supplier whose equipment design specifications and manufacturing standards meet generally accepted standards prevalent in the aquaculture industry NET

Net pen structure design, specification, and installation should be commensurate with the prevailing conditions and capable of withstanding the normal maximum weather and sea conditions.

10) Install jump nets to prevent aquatic animals from jumping out of the primary containment net NET

Jump nets should be an integral part of the primary containment net or joined to it in a fashion that prevents aquatic animal escape between the primary net and the jump net. Jump nets should be of a height appropriate to the jumping ability and size of aquatic animals they are containing. In areas with extreme winters, cages may sink slightly due to ice loads from freezing spray. This is a temporary condition that abates as the ice melts during submergence. In areas where winter icing occurs regularly, bird nets should be exchanged for winter cover nets. These nets should be constructed of netting designed to withstand the rigors of icing and with mesh sizes appropriate to contain the aquatic animal size being reared.

11) Secure nets to the appropriate attachment point, such that the attachment bears the strain and not the handrail of the cage NET

Net weights, when used for net tensioning, should be installed in a manner to prevent chafing. A second layer of net should be added one foot above and below wear points. The use of net weights should be encouraged when strong currents or tides are present at the net pen site.

12) Develop a preventative maintenance program for nets NET

The program should have the ability to track individual nets, and schedule and document regular maintenance and testing. Nets that fail testing standards should be retired and disposed of properly. An example log for recording maintenance is available in Appendix P.

13) Mooring system designs should be compatible with the net pen systems they secure NET

Mooring systems should be installed in consultation with the net pen system manufacturer or supplier. Mooring system design, specification and installation should be commensurate with the prevailing conditions of the site and be capable of withstanding the normal maximum conditions likely to occur at a site.

14) Regularly inspect and adjust mooring systems as needed NET

Rigging tension should be maintained to installation standards. New components should undergo their first inspection no later than 2 years after deployment. A diver or remote camera should regularly visually inspect subsurface mooring components. Special attention should be given to connectors and rope/chain interfaces. Chafe points should be identified and subject to more frequent inspection and removal of marine growth. With the exception of rock pin anchors, mooring systems should be hauled out of the water for a visual inspection of all components at least every 6 years. When considering what inspection method to employ, net pen operators should consider the relative risks and benefits associated with the inspection method. On sites frequently exposed to severe weather or where it is difficult to set anchors, breaking out anchors for visual, above-water inspection may represent a greater risk for mooring failure than regular underwater inspections. An example log for recording maintenance is available in Appendix P.

15) Shackles used in mooring systems should be either safety shackles, wire-tied, or welded to prevent pin drop-out NET

16) Develop a preventative maintenance program for net pen and mooring systems NET

The program should monitor maintenance of individual cages, and schedule and document regular maintenance, the nature of the maintenance, date conducted, any supporting documentation for new materials used, and who conducted the maintenance. An example log for documenting routine inspections and maintenance is available in Appendix P.

17) Use bird nets (where appropriate) to cover net cages to reduce any impacts due to bird predation; bird nets should be constructed using appropriate materials and mesh sizes designed to reduce the risk of bird entanglement NET

Contact manufacturers and suppliers of aquaculture netting for more information.

18) Develop a Standard Operating Procedure (SOP) for all routine vessel operations NET

Vessel operations around a net pen site can damage nets or the structures. All vessel operators should receive appropriate training in the operation of the vessel. The SOP should minimize the risk of damaging nets and/or mooring system components with the propeller of the vessel. When mooring barges on a permanent or semi-permanent basis, local current and wind patterns should be considered. The mooring location should be selected so that in the event of a vessel breaking free of its moorings the chance of the vessel impacting a net pen system is minimized.

Chapter 12: Record-keeping for Flow-through, Recirculating, and Net Pen Facilities

Record-keeping

Good record-keeping is the hallmark of a well-operated aquatic animal production facility. Keeping records can help a facility run more efficiently and cleanly.

Examples of Record-keeping Practices

1) Develop a record-keeping system

Records, such as feeding, chemical use, water quality, serious weather conditions, aquatic animal inventory, and aquatic animal culture operations facilitate improvements in the overall efficiency of a facility.

Record-keeping is a basic business practice and is applicable to all facilities. If a facility already has record-keeping structures in place, the existing structures can be directly used or easily adapted to incorporate any additional record-keeping requirements of the CAAP ELGs.

Record keeping is a simple, easily implemented, and cost effective management tool. Complete, well-organized records can help ensure proper maintenance of facilities and equipment and can aid in determining the causes of required repairs to help prevent future foreseeable disasters.

The following are important points to remember when performing record-keeping:

- Records must be updated regularly.
- Personnel completing and maintaining records must be trained to update records correctly.

- Records need to be readily accessible.
- Records containing any confidential information must be secured (enforcement staff can still have access to these records).

The key to maintaining records is continual updating. Ensure that new information, such as inspections of your production systems, is added to existing records as it becomes available. In addition, update records if there are changes to the number and location of discharge points, or material storage procedures. You should maintain records for at least five years from the date of sample observation or action. Some simple techniques used to accurately document and report results include:

- Forms and logs.
- Field notebooks.
- Timed and dated photographs.
- Videotapes.
- Drawings and maps.
- Computer spreadsheets and database programs.

Paper copies of records should be maintained for archival purposes; computerized record-keeping tools can be used for trend analysis and forecasting. Records should be reviewed periodically to determine if they are useful and to provide insight into opportunities for improvement of CAAP facility operation.

EPA encourages the use of existing record-keeping systems (if available at your facility) to meet the record-keeping requirements of the CAAP ELGs. However, if you need examples of forms and logs that help fulfill record-keeping requirements of the CAAP ELGs, refer to the appendices below. Some appendices contain forms and logs for activities that are not required by the ELGs, but that can be used to show your facility has met other requirements of the ELGs (e.g., showing your facility performed employee training).

- Appendix M: General Reporting Forms (*meets the CAAP ELGs general reporting requirements for INADs and extralabel drug use; failure or damage to containment systems; and spills of drugs, pesticides, and feed*)
- Appendix N: Feed Conversion Ratios Log (*may be used to track feeding and to calculate FCRs; meets the CAAP ELGs record-keeping requirements for solids control*)
- Appendix O: Spills and Leaks Log (*may be used to keep track of spills*)
- Appendix P: Inspection and Maintenance Logs (*may be used to keep track of when you perform maintenance and cleaning at your facility; meets the CAAP ELGs record-keeping requirements for maintenance*)
- Appendix Q: Cleaning Log (*may be used to document cleaning of your production systems and/or wastewater treatment systems; meets the CAAP ELGs record-keeping requirements for cleaning*)
- Appendix R: Record-keeping Checklist (*may be used to make sure you have met the record-keeping requirements of the CAAP ELGs*)
- Appendix S: Employee Training Log (*may be used to track employee training*)

- Appendix T: Carcass Removal Log (*may be used to keep track of the number of carcasses removed and disposal methods for the carcasses*)

2) Develop a record-keeping system for spills FT RAS NET

EPA requires you to report spills when they occur. EPA encourages you to keep track of spills at your facility. Records of past spills contain useful information (e.g., what practices worked best for a given magnitude and type of spill) for improving practices to prevent future spills. Typical items that should be recorded include results of routine inspections, and reported spills, leaks, or other discharges. Records should include:

- The date, exact place, and time of material inventories, site inspections, sampling observations, etc.
- Names of inspector(s) and sampler(s).
- If applicable, analytical information, including date(s) and time(s) analyses were performed or initiated, analytical techniques or methods used, the analysts' names, analytical results, and quality assurance/quality control results of such analyses.
- The date, time, exact location, and a complete characterization of significant observations, including spills or leaks.
- Notes indicating the reasons for any exceptions to standard record keeping procedures.

Refer to Appendix O for an example log to track spills at your facility.

Chapter 13: Perform Training for Flow-through, Recirculating, and Net Pen Facilities

Training

The CAAP ELGs require facilities to train all relevant personnel in spill prevention and how to respond in the event of a spill to ensure proper clean-up, and disposal of spilled materials. Facilities are also required to train personnel in the following areas:

- Operation and cleaning of production systems.

- Operation and cleaning of wastewater treatment systems.

Examples of Training Practices

1) Develop and implement an employee-training program to train relevant personnel in spill prevention and response

Employee training programs can be established to train employees how to prevent and respond to spills. Employee training programs should instill all personnel with a thorough understanding of the facility's Spill Response and Prevention Plan, including BMPs, practices for preventing spills, and procedures for responding quickly and properly to spills.

Employees can be taught through posters, employee meetings, courses, signs, and bulletin boards about spill prevention and response. Facilities may also use "in-field training" programs, where they show employees specific areas of the facility where potential spills could occur, followed by a discussion of site-specific BMPs providing solutions to spill prevention and response. Trained personnel can provide discussion to other staff within the facility.

Advantages of an employee-training program are that the program can be a low-cost and easily implementable procedure for addressing spills at aquaculture facilities. The program can be standardized and repeated as necessary, both to train new employees and to keep its objectives fresh in the minds of already trained employees. A training program is also flexible and can be adapted as a facility's management needs change over time.

Specific design criteria for implementing an employee-training program include:

- Ensuring strong commitment and periodic input from senior management.
- Communicating frequently to ensure adequate understanding of goals and objectives.
- Using experience from past spills to prevent future spills.
- Making employees aware of BMP monitoring and spill reporting procedures.
- Developing operating manuals and standard procedures.
- Implementing spill drills.

An employee-training program should be an on-going, yearly process. A sample employee training log that can be used to track employee-training programs is available in Appendix S of this document. Refer to Chapter 10 in this guidance for more specific information about developing a Spill Response and Prevention Plan.

2) Develop and implement an employee-training program to train relevant personnel in proper operation and cleaning of production and wastewater treatment systems, including feeding practices and proper use of equipment[1]

Employee training programs can be established to train employees how to properly operate and clean production and wastewater treatment systems (only flow-through and recirculating systems must train employees for operating and cleaning wastewater treatment systems), including feeding procedures and proper use of equipment.

Employees can be taught through posters, employee meetings, courses, signs, and bulletin boards about properly operating and cleaning production systems and wastewater treatment systems at your facility.

General guidance for properly operating and cleaning some of the components found in CAAP systems is available throughout the remainder of this chapter.

3) Properly operate flow-through and recirculating production systems

To properly operate production systems at your facility to reduce solids, identify what practices reduce solids (based on your facility's unique design characteristics), and

---

[1] Net pen systems are not required to train personnel in proper operation and cleaning of wastewater treatment systems.

maintain those practices. For example, maintain minimum flows to system components where required to ensure the system is self-cleaning.

Examples of other practices you can do to properly operate your systems so solids are reduced include the following:

- Avoid short-circuiting flows in the quiescent zones.
- Ensure that drainpipes and dam boards are working properly.
- Clear screens in raceways of debris.
- Do not exceed the carrying capacity of your system.
- Design and implement a feed management program.

4) Properly clean flow-through and recirculating production systems

When cleaning raceways or tanks at your facility make sure you do the following:

- Send cleaning water to a treatment system, such as an offline settling basins or full-flow settling basins.
- Clean raceways or tanks as frequently as necessary.

5) Properly clean nets

The regular cleaning of production nets helps to ensure a constant flow of water through the production area of the net pen. As the net pen sits in the culture area, marine organisms attach and grow on the nets. These organisms reduce the area of the openings. The reduction in area reduces the water flow through the net pen and the amount of dissolved oxygen available, and it increases the buildup of metabolic waste.

The following practices will help facilities to clean their nets, while minimizing the impact of this practice on the environment:

- Minimize the concentration of net-fouling organisms that are discharged during events such as changing and cleaning nets.
- Remove fouled nets, transport ashore, air dry, and clean with pressure washers, if necessary. Avoid discharges of cleaning water or net-fouling organisms to open waters.
- Avoid discharges of chemicals used to clean nets or other gear in open waters.
- Do not use materials containing or treated with tributyltin.

6) Properly operate and clean quiescent zones FT

The following practices may help you to properly operate your quiescent zone to reduce solids:

- Ensure that the turbulence is reduced (for example, preventing short circuiting by ensuring drains and outlets are operating as designed) enough so the solids will settle in the quiescent zones.
- Prevent fish from entering quiescent zones by maintaining the integrity of the screen that separates the raceways and quiescent zones.

The following guidance for cleaning quiescent zones is based on the *Idaho Waste Management Guidelines for Aquaculture Facilities* (IDEQ, n.d.).

- Settled solids should be removed regularly so they cannot become entrained in the wastewater flow and

contribute to the pollutant loadings of the facility. Two operational factors associated with operating quiescent zones are (1) the necessity to clean the screens, and (2) the regular removal of collected solids from the quiescent zones.

- Quiescent zones should be cleaned as frequently as possible, in most cases, at least once every 2 weeks.
- Screens separating the rearing area from the quiescent zone should be cleaned daily to promote laminar flow in the settling area.

7) Properly clean and operate sedimentation basins FT RAS

Solids must be removed at proper intervals to ensure the designed removal efficiencies of the sedimentation basin. For both off-line settling (OLS) and full-flow settling (FFS) basins, IDEQ recommends a minimum harvest frequency of every 6 months. Infrequent harvests could result in the breakdown of solids and the release of dissolved nutrients into the receiving waters.

For FFS basins, some facilities might batch crop their fish so that they can all be harvested at the same time. Then solids can be harvested from the FFS basins when the facility is empty (IDEQ, n.d.).

System operators should attempt to minimize the breakdown of particles (into smaller sizes) to maintain or increase the efficiency of sedimentation basins.

The following practices may be used to properly operate your sedimentation basin to reduce solids and are based on the *Idaho Waste Management Guidelines for Aquaculture Facilities* (IDEQ, n.d.):

- Regularly check the depth of collected solids and clean out the basin when the sediment depth exceeds 50% of the design depth.
- Check pipes and basin walls for cracks and other damage.
- Check for solids "caking" around the basin drain structure to ensure proper draining of treated effluent.
- Check area around the outfall for signs of erosion and repair any damage.
- Check outlet pipes for clogging.

8) Properly clean and operate microscreen filters FT   RAS

Filters require cleaning to remove trapped particles. Sprayers are used to remove collected particles and to provide additional filter cleaning. Filters may also be cleaned using a periodic rinse cycle with a heated solution.

The following practices may be used to properly operate your microscreen filters to reduce solids and are based on the *Idaho Waste Management Guidelines for Aquaculture Facilities* (IDEQ, n.d.):

- Regularly check for normal operation of the filter unit.
- Inspect all moving parts for proper operation.
- Refer to the manufacturer's operation and maintenance manual for specific details.
- Check for wear or holes in miscroscreens.
- Lubricate bearings according to manufacturer's recommended schedules.
- Check for proper operation of wash pump and cleaning nozzles.

Chapter 14: Feed Management for Net Pen Facilities

Feed Management

Waste feed and feces constitute the major portion of the wastes generated by net pens. However, because net pens operate in high-energy, open water environments (where they are exposed to currents, waves, and storms), the concentration and collection of wastes is difficult.

The most effective way to reduce the discharge of solids from net pen systems is effective feed management. Effective feed management is based on two components: waste reduction and optimal feed conversion ratio. Waste reduction focuses on ensuring that feed used by the farm is not lost or discharged prior to intake by the aquatic animal. Optimal conversion focuses on ensuring that all feed intake offered to the aquatic animal is actually consumed and optimally digested and used by the aquatic animal.

Documenting efficient feed management for EPA may be achieved by describing the following in your BMP plan:

- Feed methods used to minimize solids production.
- Modifications made to feed quantities as fish production changes (e.g., size, health of fish).
- Feed handling methods used to reduce generation of fines.
- Feed formulations information for each life-history stage of fish reared.

Examples of Feed Management Practices

1) Calculate feed conversion ratios by using feed and aquatic animal biomass inventory tracking systems

Calculation of feed conversion ratios (FCRs) is an essential function on all net pen farms. Monitoring long- and short-term changes in feed conversion ratios allows farmers to quickly identify significant changes in feed consumption and waste production rates. Refer to Appendix N for an example log to calculate and track FCRs.

2) In cooperation with feed manufacturers, seek to minimize nutrient and solids discharges through optimization of feed formulations

Feeds should be formulated for optimum feed conversion ratios and retention of protein (nitrogen) and phosphorus. Feed formulations should consider numerous factors including, pellet stability, digestibility, palatability, sinking rates, energy levels, moisture content, ingredient quality and the nutritional requirements of the species being grown. Feeds should be formulated and manufactured using high-quality ingredients. Feed ingredients should have high dry matter and protein apparent digestibility coefficients. Formulations should be designed to enhance nitrogen and phosphorus retention efficiency, and reduce metabolic waste output. Feeds should contain sufficient dietary energy to spare dietary protein (amino acids) for tissue synthesis. Feeds should be water stable for sufficient periods such that pellets remain intact until eaten by fish. Questions regarding feed formulations should be referred to a qualified fish nutritionist or feed manufacturer.

3) Experiment with feed formulations designed to reduce the total environmental impact of the feed

If experimental formulations that use alternative protein and lipid sources are tried, care should be taken to ensure that digestibility is not decreased and the nutritional needs of the species being cultured are met. Farmers should be careful that alternate formulations do not increase feed conversion ratios, decrease fish growth, and result in increased fecal waste.

4) Use efficient feeding practices [NET]

Feed can be delivered by hand, demand feeders, automatic feeders, or by mechanical feeders. Regardless of the delivery method or system, the amount of feed offered should optimize the balance between maximum growth and maximum feed conversion efficiency. The appropriate quantity and type of feed for a given species is influenced by aquatic animal size, water temperature, dissolved oxygen levels, health status, reproductive status, and management goals. Feed particle size should be appropriate for the size of aquatic animals being fed. Feeding behavior should be observed to monitor feed utilization and evaluate health status.

5) Check feeding equipment to ensure efficient operation [NET]

Improperly adjusted or malfunctioning feeding equipment can over-feed or under-

feed fish and reduce feed and production efficiency.

Figure 14.1. Automatic feeder for net pens

Figure 14.2. Feed handling for net pens

6) Reduce fish stress and optimize culture conditions to reduce FCRs [NET]

Facilities can reduce fish stress by avoiding overcrowding in production systems and maintaining and cleaning net pens to make sure adequate water can move through the nets. Remember to properly clean your nets to avoid harm to the environment.

7) Conduct employee training in fish husbandry and feeding methods to ensure that workers have adequate training to optimize FCRs [NET]

Additional information about performing training is available in "Chapter 13: Perform Training for Flow-through, Recirculating, and Net Pen Facilities."

8) Wherever practical, use monitoring technologies such as video, "lift-ups," or digital scanning sonar sensors to monitor feed consumption and reduce feed waste [NET]

If automated feeding systems are used, fish monitoring systems should, if possible, be actively linked to feeding control systems to provide direct control feedback to reduce feed wastage. Even if monitoring systems are employed, active monitoring by farm operators should also occur to ensure that all systems are functioning properly and aquatic animals are behaving and feeding normally.

9) If water depths and currents allow, regularly examine the bottom under net pens and cages [NET]

To prevent benthic impacts from occurring, close attention should be paid to the presence of any waste feed and how the benthic environment appears to be assimilating the nutrient load. Regular inspections by divers or video cameras can alert farm operators to potential problems before they become unmanageable. Also, use information collected by third parties or regulators to adjust management practices if necessary.

Chapter 15: Waste Collection and Disposal, Transport or Harvest Discharge, and Carcass Removal for Net Pen Facilities [NET]

The CAAP ELGs require net pen facilities to collect and properly dispose of solid waste (such as feed bags, packaging materials, rope, or netting). In addition, net pen facilities are required to minimize the discharge associated with harvest and transport, particularly blood, viscera, or animal carcasses. These facilities should also prevent the discharge of animal mortalities by properly removing and disposing of carcasses. The following describes practices that may be used to achieve these requirements.

Waste Collection and Disposal

The CAAP ELGs require facilities to collect, return to shore, and properly dispose of all feed bags, packaging materials, waste rope, and netting.

Examples of Waste Collection and Disposal Practices

1) Conduct a systematic review of your operation; a waste management plan can be used to effectively manage, use, and dispose of wastes generated during production; the plan identifies all wastes generated on a site or from a facility [NET]

Waste management plans clearly identify all wastes generated on a site and classify them with respect to any risks associated with their collection and appropriate disposal. The waste management plan may be designed to minimize the generation of waste while recognizing the practical challenges associated with marine operations.

Waste management plans encourage recycling of waste except when human or animal health may be compromised. In these cases, a clear containment and disposal method may be outlined. These methods and actions may be designed to minimize any human or fish health risks associated with the waste. Waste management plans may address feed bags, packaging materials, waste rope, and netting. Other wastes include aquatic animal mortalities and chemical/fuel spills. These substances are addressed in the next section on "Carcass Removal" and in "Chapter 13: Perform Training for Flow-through, Recirculating, and Net Pen Facilities," respectively.

2) Avoid the discharge of substances associated with in-place pressure washing of nets into the waters of the United States [NET]

Whenever possible, use gear and production strategies that minimize or eliminate the need for on-site wash down and rinsing to reduce biofouling. The use of air-drying, mechanical, biological, and other non-chemical procedures to control net fouling are strongly encouraged. In some areas with high flushing rates or great depth, in-place net washing may be acceptable. In areas with high fouling rates, treatment of nets with anti-fouling compounds permitted by EPA may represent a lower environmental risk than frequent net washing.

3) Collect, return to shore, and properly dispose of all feed bags, packaging materials, waste rope, and netting (using methods approved by appropriate regulatory authorities); recycling is strongly encouraged NET

4) Be proactive about minimizing all types of solid waste generation NET

Facilities should review their operations and consider whether there are alternative practices that help reduce the use of materials that generate solid waste. For example, consider the use of packaging and materials handling methods that reduce total packaging needs.

Transport or Harvest Discharge

Facilities should properly dispose of transport or harvest discharge (e.g., viscera, blood) when aquatic animals are harvested.

Examples of Transport or Harvest Discharge Practices

1) Design and operate harvest procedures and equipment in a fashion that reduces any associated discharges; harvest and post-harvest vessel and equipment clean up procedures should minimize any wastes discharged overboard NET

2) Collect and properly dispose of any processing and harvesting waste NET

Facilities should dispose of processing or harvesting waste in a manner that prevents it from entering into waters of the United States.

It may be useful to keep a general operations log to track activities at your facility concerning waste disposal, transport or harvest discharge, and carcass removal. For example:

- *9/15/04: hauled feed bags and waste rope to shore; disposed of these materials in a dumpster.*
- *10/4/04: transported aquatic animals (no water spilled).*

Carcass Removal

Proper aquatic animal health management is the best method of managing mortalities in net pens and cages. Optimizing aquatic animal health will reduce the need to deal with dead fish. Even under optimal conditions some mortalities can occur. Net pens should contain and collect any mortalities that may occur. This facilitates the close monitoring of mortality rates and their timely removal. Severe weather may temporarily prevent mortality removal. Remove mortalities as soon as weather permits. Keeping records of severe weather days is recommended.

An example log for tracking carcass removal and disposal is available in Appendix T. This log could be useful for facilities tracking aquatic animal mortalities and in subtracting out mortalities from calculations for feed conversion ratios.

Examples of Carcass Removal Practices

1) Weather permitting, regularly and frequently collect mortalities to prevent their discharge to waters of the United States NET

When collecting and removing mortalities, use methods that do not stress remaining

---

animals, jeopardize worker safety, or compromise biosecurity. Mortalities should only be stored and transported in closed containers with tight fitting lids. Mortalities should be returned to shore and disposed of properly, using methods approved by appropriate regulatory authorities. Facilities may want to consider practices such as composting as a method to treat mortalities.

As part of your facility's BMP plan, outline the process for removing and properly disposing of carcasses from your facility.

2) Proactively manage your aquatic animal stocks to optimize animal health NET

References and Resources

References

IDEQ. n.d. Idaho Waste Management Guidelines for Aquaculture Operations. Idaho Department of Environmental Quality. <http://www.deq.state.id.us/water/prog_issues/waste_water/pollutant_trading/aquaculture_guidelines.pdf>. Accessed December 2004.

Tucker, C., S. Belle, C. Boyd, G. Fornshell, J. Hargreaves, S. LaPatra, S. Summerfelt, and P. Zajicek. 2003. *Best Management Practices for Flow-Through, Net-Pen, Recirculating, and Pond Aquaculture Systems*.

Additional Resources

Lawson, T.B. 1995. *Fundamentals of Aquacultural Engineering*. pp. 48-57. Chapman & Hall, NY.

Metcalf and Eddy, Inc. 1991. *Wastewater Engineering: Treatment and Disposal*, 3d ed., revised by G. Tchobanoglous and F. Burton. McGraw Hill, Inc., NY.

Soderberg, R.W. 1995. *Flowing Water Fish Culture*. Lewis Publishers, Boca Raton, FL.

Tomasso, J., ed. 2002. *Aquaculture and the Environment in the United States*. U.S. Aquaculture Society, A Chapter of the World Aquaculture Society, Baton Rouge, LA.

USEPA. 1996. *NPDES Permit Writers' Manual*. EPA-833-B-96-003. U.S Environmental Protection Agency, Office of Water, Washington, DC. <http://cfpub.epa.gov/npdes/pkeyword.cfm?keywords=permit+writers&program_id=0>. Accessed December 2004.

USEPA. 2004. *Economic and Environmental Benefit Analysis of the Final Effluent Limitations Guidelines and Standards for the Concentrated Aquatic Animal Production Point Source Category*. EPA 821-R-04-013, U.S. Environmental Protection Agency, Office of Water, Washington, DC.

USEPA. 2004. *Technical Development Document for the Final Effluent Limitations Guidelines and New Source Performance Standards for the Concentrated Aquatic Animal Production Point Source Category (Revised August 2004)*. EPA 821-R-04-012, U.S. Environmental Protection Agency, Office of Water, Washington, DC. <http://epa.gov/guide/aquaculture/ >. Accessed December 2004.

USEPA. 2004. *Website for the Aquatic Animal Production Industry Effluent Guidelines*. <http://epa.gov/guide/aquaculture/>. Accessed December 2004.

Wedemeyer, G.A. ed. 2001. *Fish Hatchery Management*, 2d ed., American Fisheries Society, Bethesda, MD.

Wheaton, F.W. 1977. *Aquacultural Engineering*. pp. 643-679. John Wiley and Sons, Inc., NY.

Appendix A

State Permitting Authorities/Departments of Environmental Protection

State Permitting Authorities/Departments of Environmental Protection

Alabama Alabama Dept. of Environmental Management· Permit and Services Division Post Office Box 301463 Montgomery, Alabama 36130-1463 (334) 271-7714 http://www.adem.state.al.us	**Colorado** Colorado Department of Public Health and Environment, Water Quality Control Division 4300 Cherry Creek Drive South Denver, CO 80246 (303) 692-3500 http://www.cdphe.state.co.us/wq/wqhom.asp
Alaska U.S. Environmental Protection Agency, Region 10 1200 6th Avenue Seattle, WA 98101-1128 (206) 553-1200 or (800) 424-4EPA http://yosemite.epa.gov/R10/WATER.NSF	**Connecticut** Connecticut Dept. of Environmental Protection, Bureau of Water Management, Permitting, Enforcement and Remediation Division 79 Elm Street Hartford, CT 06106 (806) 424-3018 http://dep.state.ct.us/wtr/prgactiv.htm
American Samoa No information found	**Delaware** Delaware Dept. of Natural Resources and Environmental Control, Division of Water Resources 89 Kings Highway Dover, DE 19901 (302) 739-4860 http://www.dnrec.state.de.us/DNREC2000/ WaterResources.asp
Arizona Arizona Department of Environmental Quality, Office of Water Quality 1110 West Washington Street Phoenix, AZ 85007 *Phone:* (602) 771-2300 http://www.adeq.state.az.us	**Florida** Florida Department of Environmental Protection 3900 Commonwealth Boulevard M.S. 49 Tallahassee, Florida 32399 (850)-245-2118 http://www.dep.state.fl.us Florida Dept. of Agriculture and Consumer Services Division of Aquaculture 1203 Governors Square Boulevard, Fifth Floor Tallahassee, FL 32301 (850) 488-4033 http://www.FloridaAquaculture.com
Arkansas Arkansas Department of Environmental Quality 8001 National Drive P.O. Box 8913 Little Rock, AR 72219 (50l) 682-0744 http://www.adeq.state.ar.us	**Georgia** Georgia Department of Natural Resources Environmental Protection Division 2 Martin Luther King Jr. Dr., Suite 1152 East Tower Atlanta, GA 30334 (404-657-5947) or (888-373-5947) http://www.gaepd.org
California California State Water Resources Control Board, Division of Water Quality 1001 I Street, 15th Floor Sacramento, CA 95814 (916) 341-5250 http://www.swrcb.ca.gov	**Guam** No information found

Hawaii Hawaii Dept. of Health, Environmental Health 1250 Punchbowl Street Honolulu, HI 96813 (808) 586-4400 http://www.hawaii.gov/health Mailing Address: P.O. Box 3378, Honolulu, HI 96801	**Louisiana** Louisiana Department of Environmental Quality Office of Environmental Services P. O. Box 4313 Baton Rouge, LA 70821-4313 (225) 219-3181 http://www.deq.state.la.us
Idaho U.S. Environmental Protection Agency, Region 10 1200 6th Avenue Seattle, WA 98101-1128 (206) 553-1200 or (800) 424-4EPA http://yosemite.epa.gov/R10/WATER.NSF	**Maine** Maine Department of Environmental Protection Bureau of Land and Water Quality 17 State House Station Augusta, ME 04333 (207) 287-7688 or (800) 452-1942 http://www.state.me.us/dep/blwq/index.htm
Illinois Illinois EPA, Bureau of Water 1021 North Grand Ave. East, P.O. Box 19276 Springfield, IL 62794-9276 (217) 782-3362 http://www.epa.state.il.us/water	**Maryland** Maryland Department of the Environment 1800 Washington Blvd. Baltimore, MD 21230 (410) 537-3000 http://www.mde.state.md.us
Indiana Indiana Dept. of Environmental Management Office of Water Quality 100 North Senate Avenue, P.O. Box 6015 Indianapolis, IN 46206-6015 (317) 232-8603 or (800) 451-6027 (toll free: IN) http://www.in.gov/idem/water	**Massachusetts** U.S. Environmental Protection Agency, Region 1 One Congress Street, Suite 1100 Boston, MA 02114-2023 (617) 918-1111 or (888) 372-7341 (New England states) http://www.epa.gov/region1/npdes/mass.html
Iowa Iowa Department of Natural Resources 502 E. 9th Street, Henry A. Wallace State Office Bldg. Des Moines, IA 50319-0034 (515) 281-5918 http://www.iowadnr.com	**Michigan** Michigan Department of Environmental Quality Water Bureau Constitution Hall, 525 West Allegan St, P.O. Box 30473 Lansing, MI 48909-7973 (517) 373-7917 http://www.michigan.gov/deq
Kansas Kansas Department of Health and Environment Bureau of Water 1000 Southwest Jackson Street, Suite 420 Topeka, KS 66612-1367 (785) 296-5500 http://www.kdhe.state.ks.us/water/index.html	**Minnesota** Minnesota Pollution Control Agency 520 Lafayette Road St. Paul, MN 55155-4194 651-297-2274 or 800-646-6247 http://www.pca.state.mn.us
Kentucky Kentucky Dept. for Environmental Protection Division of Water 14 Reilly Road Frankfort, KY 40601 (502) 564-3410 http://www.water.ky.gov	**Mississippi** Mississippi Dept. of Environmental Quality, Office of Pollution Control, Environmental Permits Division P.O. Box 10385 Jackson, MS 39289-0385 (601) 961-5171 or (888) 786-0661 http://www.deq.state.ms.us/MDEQ.nsf/page/ Main_Home?OpenDocument

Missouri Missouri Dept. of Natural Resources, Water Pollution Control Branch, Permits Section P.O. Box 176 Jefferson City, MO 65102-0176 (573) 751 –3443 or (800) 361-4827 http://www.dnr.mo.gov/env/wpp/permits/index.html	**New Mexico** U.S. Environmental Protection Agency, Region 6 1445 Ross Avenue Dallas, TX 75202 (215) 665-6444 http://www.epa.gov/region6
Montana Montana Dept. of Environmental Quality, Permitting and Compliance Division, Water Protection Bureau 1520 E. Sixth Avenue P.O. Box 200901 Helena, MT 59620-0901 *Phone:* (406) 444-2544 http://www.deq.state.mt.us/pcd/wpb/index.asp	**New York** New York Department of Environmental Conservation, Division of Water 625 Broadway Albany, NY 12233 (518) 402-8111 http://www.dec.state.ny.us/website/dow
Nebraska Nebraska Department of Environmental Quality 1200 N Street, Suite 400, P.O. Box 98922 Lincoln, NE 68509 (402) 471-2186 http://www.deq.state.ne.us	**North Carolina** North Carolina Department of Environment & Natural Resources, Division of Water Quality 1617 Mail Service Center Raleigh, NC 27699-1617 (919) 733-7015 http://www.enr.state.nc.us
Nevada Nevada Division of Environmental Protection, Bureau of Water Pollution Control 333 West Nye Lane, Suite 138 Carson City, NV 89706-0851 (775) 687-9418 http://ndep.nv.gov/bwpc/bwpc01.htm	**North Dakota** North Dakota Health Department Division of Water Quality 1200 Missouri Avenue P.O. Box 5520 Bismarck, ND 58502-5520 (701) 328-5210 http://www.health.state.nd.us/wq
New Hampshire U.S. Environmental Protection Agency, Region 1 One Congress Street, Suite 1100 Boston, MA 02114-2023 (617) 918-1111 or (888) 372-7341 (New England states) http://www.epa.gov/region1/npdes/newhampshire.html	**Ohio** Ohio Environmental Protection Agency Division of Surface Water Lazarus Government Center 122 South Front Street P.O. Box 1049 Columbus, OH 43216 (614) 644-2021 http://web.epa.state.oh.us/dsw
New Jersey New Jersey Department of Environmental Protection, Bureau of Nonpoint Pollution Control 401 East State Street, P.O. Box 29 Trenton, NJ 08625-0029 (609) 633-7021 http://www.state.nj.us/dep/dwq/nonpoint.htm	**Oklahoma** U.S. Environmental Protection Agency, Region 6 1445 Ross Avenue Dallas, TX 75202 (215) 665-6444 http://www.epa.gov/region6

Oregon Oregon Department of Environmental Quality 811 SW Sixth Avenue Portland, OR 97204-1390 (503) 229-5696 or (800) 452-4011 (in Oregon) http://www.deq.state.or.us	**Tennessee** Tennessee Department of Environment and Conservation, Division of Water Pollution Control 401 Church Street L&C Tower 21st Floor Nashville, TN 37243 (888) 891-8332 http://www.state.tn.us/environment/wpc
Pennsylvania Pennsylvania Dept. of Environmental Protection Office of Water Management 16th Floor, Rachel Carson State Office Building P.O. Box 2063 Harrisburg, PA 17105-2063 (717) 787-4686 http://www.dep.state.pa.us/dep/deputate/watermgt/watermgt.htm	**Texas** Texas Commission on Environmental Quality 1700 North Congress Avenue P.O. Box 13087 Austin, TX 78711-3087 (512) 239-1000 http://www.tceq.state.tx.us/AC/nav/permits/water_qual.html
Puerto Rico U.S. Environmental Protection Agency, Region 2 290 Broadway New York, NY 10007-1866 (212) 637-5000 http://www.epa.gov/Region2/water/wpb/npdes.htm	**Utah** Utah Department of Environmental Quality Division of Water Quality 288 North 1460 West Cannon Building, 3rd Floor P.O. Box 144870 Salt Lake City, UT 84114-4870 (801) 538-6146 http://waterquality.utah.gov
Rhode Island Rhode Island Department of Environmental Management, Office of Water Resources 235 Promenade Street Providence, RI 02908 (401) 222-6800 http://www.state.ri.us/dem/programs/benviron/water/index.htm	**Vermont** Vermont Department of Environmental Conservation Wastewater Management Division 103 South Main Street Sewing Bldg. Waterbury, VT 05671-0405 (802) 241-3822 http://www.anr.state.vt.us/dec/ww/wwmd.cfm
South Carolina South Carolina Department of Health & Environmental Control 2600 Bull Street Columbia, SC 29201 (803) 898-3432 http://www.scdhec.net	**Virginia** Virginia Department of Environmental Quality 629 East Main Street P.O. Box 10009 Richmond, VA 23240-0009 (804) 698-4000 or 1-800-592-5482 (in Virginia) http://www.deq.state.va.us
South Dakota South Dakota Department of Environment and Natural Resources, Surface Water Quality Program Joe Foss Building 523 East Capitol Avenue Pierre, SD 57501 (605) 773-3351 http://www.state.sd.us/denr/denr.html	**Virgin Islands** U.S. Environmental Protection Agency, Region 2 290 Broadway New York, NY 10007-1866 (212) 637-5000 http://www.epa.gov/Region2/water/wpb/npdes.htm

Washington Washington Department of Ecology P.O. Box 47600 Olympia, WA 98504 (360) 407-6413 http://www.ecy.wa.gov/ecyhome.html	**Wisconsin** Wisconsin Department of Natural Resources Bureau of Wastewater Management 101 South Webster Street P.O. Box 7921 Madison, WI 53707 (608) 267-7694 http://www.dnr.state.wi.us
West Virginia West Virginia Department of Environmental Protection, Division of Water and Waste Management, Water Permitting Section 601 - 57th Street Charleston, WV 25304 (304) 926-0495 http://www.dep.state.wv.us	**Wyoming** Wyoming Department of Environmental Quality Water Quality Division 122 West 25th Street Herschler Building, 4th Floor West Cheyenne, WY 82001 (307) 777-7781 http://deq.state.wy.us/wqd

Appendix B

Natural Resources Agencies
Associated with Fisheries

Natural Resources Agencies Associated with Fisheries

Alabama Alabama Department of Conservation and Natural Resources 64 N. Union Street, Suite 468 Montgomery, Alabama 36130 (334) 242-3486 http://www.dcnr.state.al.us	**Colorado** Colorado Department of Natural Resources Division of Wildlife 1313 Sherman St., Rm. 718 Denver, CO 80203 (303) 866-3311 http://wildlife.state.co.us
Alaska Alaska Department Fish and Game P.O. Box 25526 Juneau, Alaska 99802-5526 (907) 465-4100 http://www.adfg.state.ak.us	**Connecticut** Connecticut Department of Environmental Protection Bureau of Natural Resources 79 Elm Street Hartford, CT 06106-5127 (860) 424-3010 http://dep.state.ct.us
American Samoa Department of Marine and Wildlife Resources American Samoa Government, Executive Office Building, Utulei Territory of American Samoa, Pago Pago, AS 96799 (684) 633-4456 http://www.asg-gov.net http://www.asg-gov.net/ MARINE%20&%20WILDLIFE%20RESOURCES.htm	**Delaware** Delaware Department of Natural Resources and Environmental Control Division of Fish and Wildlife 89 Kings Hwy. Dover, DE 19901 (302) 739-3441 http://www.dnrec.state.de.us/dnrec2000/index.asp
Arizona Arizona Game and Fish 2221 W. Greenway Rd. Phoenix, AZ 85023-4399 (602) 942-3000 http://www.gf.state.az.us	**Florida** Florida Fish and Wildlife Conservation Commission 620 South Meridian Street Tallahassee, FL 32399-1600 http://myfwc.com
Arkansas Arkansas Game & Fish Commission 2 Natural Resources Drive Little Rock, Arkansas 72205 (800) 364-4263 http://www.agfc.state.ar.us	**Georgia** Georgia Department of Natural Resources 2 Martin Luther King, Jr. Drive, S. E. Suite 1252 East Tower Atlanta, GA 30334 (404) 656-3500 http://www.gadnr.org
California California Department of Fish and Game 1416 Ninth Street Sacramento, California 95814 Phone: (916) 445-0411 (916) 445-0411 http://www.dfg.ca.gov	**Guam** Guam Department of Agriculture Division of Aquatic and Wildlife Resources 192 Dairy Road Mangilao, Guam 96923 (671) 735-3986

Hawaii State of Hawaii, Department of Land and Natural Resources Kalanimoku Bldg. 1151 Punchbowl St. Honolulu, HI 96813 (808) 587-0400 http://www.hawaii.gov/dlnr	**Louisiana** Louisiana Department of Wildlife and Fisheries 2000 Quail Drive Baton Rouge, La. 70808 (225) 342-4500 http://www.wlf.state.la.us/apps/netgear/page1.asp
Idaho Idaho Fish and Game 600 S Walnut PO Box 25 Boise, ID 83707 (208) 334-3700 http://fishandgame.idaho.gov	**Maine** Maine Department of Environmental Protection 17 State House Station Augusta, ME 04333-0017 (207) 287-7688 http://www.maine.gov/dep/index.shtml
Illinois Illinois Department of Natural Resources James R. Thompson Center 100 W. Randolph St., Suite 4-300 Chicago, IL 60601 (312) 814-2070 http://dnr.state.il.us/about/officeadd.htm	**Maryland** Maryland Department of Natural Resources 580 Taylor Avenue Tawes State Office Building Annapolis, MD 21401 http://www.dnr.state.md.us/sw_index_flash.asp
Indiana Indiana Department of Natural Resources 402 West Washington Street Indianapolis, IN 46204 http://www.in.gov/dnr	**Massachusetts** Massachusetts Division of Fisheries and Wildlife 251 Causeway Street, Suite 400 Boston MA 02114-2154 (617) 626-1590 http://www.mass.gov/dfwele/dfw/dfw_toc.htm
Iowa Iowa Department of Natural Resources 502 E. 9th Street Des Moines, IA 50319-0034 (515) 725-0275 http://www.iowadnr.com	**Michigan** Michigan Department of Natural Resources Fisheries Division P.O. Box 30446 Lansing, MI 48909 (517) 373-1280 http://www.michigan.gov/dnr
Kansas Kansas Department of Wildlife and Parks 1020 S. Kansas Ave., Suite 200 Topeka, KS 66612 (785) 296-2281 http://www.kdwp.state.ks.us	**Minnesota** Department of Natural Resources 500 Lafayette Road St. Paul, MN 55155-4040 (651) 296-6157 http://www.dnr.state.mn.us/index.html
Kentucky Kentucky Fish and Wildlife Resources #1 Game Farm Road Frankfort, KY 40601 (800) 858-1549 http://www.kdfwr.state.ky.us	**Mississippi** Mississippi Wildlife, Fisheries, and Parks 1505 Eastover Drive Jackson, MS 39211-6374 (601) 432-2400 http://www.mdwfp.com

Missouri Missouri Department of Conservation P. O. Box 176 Jefferson City, MO 65102 (800) 361-4827 http://mdc.mo.gov	**New Mexico** New Mexico Game and Fish One Wildlife Way Santa Fe, NM 87507 (505) 476-8000 http://www.wildlife.state.nm.us
Montana Montana Fish, Wildlife and Parks 1420 East 6th Avenue Helena, MT 59620-0701 (406) 444-2535 http://fwp.state.mt.us/default.html	**New York** NY State Dept. of Environmental Conservation Office of Natural Resources and Water Division of Fish, Wildlife and Marine Resources 625 Broadway Albany, NY 12233 (518) 402-8924 http://www.dec.state.ny.us/website/dfwmr/index.html
Nebraska The Nebraska Game and Park Commission 2200 N. 33rd St. Lincoln, NE 68509 (402) 471-0641 http://www.ngpc.state.ne.us	**North Carolina** North Carolina Department of Environment and Natural Resources 1601 Mail Service Center Raleigh, NC 27699-1601 (919) 733-4984 http://www.enr.state.nc.us
Nevada Nevada Department of Conservation and Natural Resources 123 W. Nye Lane, Room 230 Carson City, NV 89706-0818 (775) 687-4360 http://dcnr.nv.gov	**North Dakota** North Dakota Game and Fish Department 100 N. Bismarck Expressway Bismarck, ND 58501-5095 (701) 328-6300 http://gf.nd.gov/
New Hampshire New Hampshire Fish and Game Department 11 Hazen Drive Concord, NH 03301 (603) 271-3421 http://www.wildlife.state.nh.us	**Ohio** Ohio Department of Natural Resources Division of Wildlife 2045 Morse Road, Bldg. G Columbus, OH 43229 (614) 265-6300 http://www.dnr.state.oh.us/wildlife/default.htm
New Jersey New Jersey Department of Environmental Protection Division of Fish and Wildlife 501 E. State St., 3rd Floor Trenton, NJ 08625-0400 http://www.state.nj.us/dep/fgw	**Oklahoma** Oklahoma Department of Wildlife Conservation 1801 N. Lincoln Oklahoma City, OK 73105 (405) 521-3721 http://www.wildlifedepartment.com

Oregon Oregon Department of Fish and Wildlife 3406 Cherry Avenue NE Salem, OR 97303 (503) 947-6000 http://www.dfw.state.or.us	**Tennessee** Tennessee Wildlife Resources Agency P.O. Box 40747 Nashville, TN 37204 (615) 781-6500 http://www.state.tn.us/twra
Pennsylvania Pennsylvania Fish and Boat Commission 1601 Elmerton Avenue Harrisburg, PA 17110 (717) 705-7800 http://www.fish.state.pa.us	**Texas** Texas Parks and Wildlife Services 4200 Smith School Road Austin, TX 78744 http://www.tpwd.state.tx.us/fish
Puerto Rico Departamento de Recursos Naturales y Ambientales (DRNA) (Natural & Environmental Resources Department) P.O Box 9066600, Puerta de Tierra Station Santurce, PR 00906 (787) 724-8774 http://www.gobierno.pr/drna	**Utah** Utah Department of Natural Resources 1594 West North Temple Salt Lake City, UT 84114 http://www.water.utah.gov
Rhode Island State of Rhode Island Department of Environmental Management 4808 Tower Hill Road Wakefield, RI 02879 (401) 222-6800 http://www.state.ri.us/dem/programs/bnatres/fishwild	**Vermont** Vermont Agency of Natural Resources 103 South Main Street Center Building Waterbury, VT 05671-0301 (802) 241-3600 http://www.anr.state.vt.us
South Carolina South Carolina Department of Natural Resources Rembert C. Dennis Building 1000 Assembly Street Columbia, SC 29201 (803) 734-3886 http://www.dnr.state.sc.us	**Virginia** Virginia Department of Game and Inland Fisheries 4010 West Broad St. Richmond, VA 23230 (804) 367-1000 http://www.dgif.state.va.us
South Dakota South Dakota Game, Fish and Parks 523 E Capitol Pierre, SD 57501 (605) 773-3381 http://www.sdgfp.info/Index.htm	**Virgin Islands** U.S. Virgin Islands Department of Planning and Natural Resources Cyril E. King Airport, 2nd Floor St. Thomas, US Virgin Islands 00802 (340) 774-3320 45 Mars Hills, Frederiksted St. Croix, US Virgin Islands 00841 (340) 773-1082 http://www.dpnr.gov.vi/about.htm

Washington	**Wisconsin**
Washington Department of Fish and Wildlife	Wisconsin of Department of Natural Resources
Natural Resources Building	101 S Webster St
Olympia, WA 98501	PO Box 7921
(360) 902-2200	Madison Wisconsin 53707-7921
http://wdfw.wa.gov	http://www.dnr.state.wi.us
West Virginia	**Wyoming**
West Virginia Division of Natural Resources	Wyoming Game and Fish
Wildlife Resources	5400 Bishop Blvd.
Fisheries Management	Cheyenne, WY 82006
1900 Kanawha Boulevard, E.	(307) 777-4600
Charleston, WV 25305	http://gf.state.wy.us/fish/index.asp
(304) 558-2771	
http://www.wvdnr.gov/Fishing/Fishing.shtm	

Appendix C

Frequently Asked Questions

Frequently Asked Questions

What operations are covered under the CAAP ELGs?

The CAAP ELGs apply to direct discharges of wastewater from existing and new facilities in these two categories:

- Facilities that produce at least 100,000 pounds a year in flow-through and recirculating systems that discharge wastewater at least 30 days a year (used primarily to raise trout, salmon, hybrid striped bass, and tilapia).
- Facilities that produce at least 100,000 pounds a year in net pens or submerged cage systems (used primarily to raise salmon).

What do the CAAP ELGs require?

The rule requires that all applicable facilities:

- Develop, maintain, and certify a Best Management Practice plan that describes how the facility will meet the requirements of the regulation.
- Prevent discharge of drugs and pesticides that have been spilled and minimize discharges of excess feed.
- Regularly maintain production and wastewater treatment systems.
- Keep records on numbers and weights of animals, amounts of feed, and frequency of cleaning, inspections, maintenance, and repairs.
- Train staff to prevent and respond to spills and to properly operate and maintain production and wastewater treatment systems.
- Report the use of experimental animal drugs or drugs that are not used in accordance with label requirements.
- Report failure of or damage to a containment system.

The rule requires flow through and recirculating discharge facilities to minimize the discharge of solids such as uneaten feed, settled solids, and animal carcasses.

The rule requires open water system facilities (e.g., net pens or cages in the ocean) to:

- Employ efficient feed management strategies to allow only the least possible uneaten feed to accumulate beneath the nets.
- Properly dispose of feed bags, packaging materials, waste rope, and netting.
- Limit as much as possible wastewater discharges resulting from the transport or harvest of the animals.
- Prevent the discharge of dead animals to waters of the U.S.

Additional information about these requirements is available from Chapters 6 and 7, and Chapters 9 through 15 of this guidance.

What operations are covered under the NPDES regulation?

EPA's existing NPDES regulations define when a hatchery, fish farm, or other facility is a CAAP facility and, therefore, a point source subject to the NPDES permit program. See 40 CFR 122.24. In defining CAAP facilities, the NPDES regulations distinguish between warm water and cold water species of fish and define a CAAP facility by, among other things, the size of the operation and frequency of discharge. A facility is a CAAP facility if it meets the criteria in 40 CFR 122, Appendix C (available in Appendix D of this guide) or if it is designated as a CAAP facility by the Director on a case-by-case basis.

Most facilities falling under the definition of CAAP are either flow-through, recirculating, or net pen systems. These systems discharge continuously or discharge 30 days or more per year as defined in 40 CFR part 122.24 and are subject to permitting depending on the production level at the facility. Most pond facilities do not require permits because ponds generally discharge fewer than 30 days per year and therefore generally are not CAAP facilities, unless designated by the Director.

In general[1], you will not be covered by the NPDES regulations if you are a facility that produces less than 9,090 harvest weight kilograms (approximately 20,000 pounds) per year of cold water species or if you feed less than 2,272 kilograms (approximately 5,000 pounds) of food during the calendar month of maximum feeding (40 CFR part 122, Appendix C). The NPDES regulations also do not apply if you are a facility that produces warm water species, using closed ponds that discharges only during periods of excess runoff or if you produce less than 45,454 harvest weight kilograms (about 100,000 pounds) per year of warm water species (40 CFR part 122, Appendix C).

Facilities meeting the NPDES definition of a CAAP will still be regulated by the NPDES permit program, even if they are not subject to the ELGs.

What is the difference between NPDES and ELGs for CAAPs?

Any facility can be designated as a CAAP (whether it meets the requirements of the NPDES regulations outright or whether the Director designates the facility as a CAAP facility) and be subject to NPDES permitting requirements.

However, if the ELG applies to the CAAP facility (i.e., recirculating, flow-through, or net pen systems that annually produce more than 100,000 pounds of aquatic animals) then the facility's NPDES permit will also contain ELGs requirements specific to the system types used to produce aquatic animals at that location.

---

[1] Unless the Director designates your facility as requiring an NPDES permit.

When do the ELGs take effect?

The ELGs requirements will apply during a facility's next permit cycle (i.e., when the facility's permit is renewed).

How does EPA define a facility?

A facility is defined as all contiguous property and equipment owned, operated, or leased, or under control of the same person or entity. Each system owned, operated, leased, or under the control of the same person or entity that is not contiguous can and should be treated as separate facilities; the production threshold used in determining if a facility is a CAAP should also be applied separately.

What is annual production?

EPA defines annual production for aquatic animal production facilities as what aquatic animals leave the facility on an annual basis. Check with your permitting authority to verify how they define production.

What if I have more than one type of system (i.e., recirculating and net pen) at my facility?

If you have more than one type of regulated system (flow-through, recirculating, or net pen) at your facility (and the total annual production at any one of the systems is 100,000 pounds or more), you must comply with the different requirements for each system type. For example, if you have a recirculating system and net pens at your facility, you will need to comply with the appropriate ELGs requirements for both recirculating systems and net pens. For more information about different system types and meeting the ELGs' production threshold, refer to Chapter 3.

If you have other system types in addition to those regulated by the ELGs, such as ponds or shellfish hatcheries, those system types are not regulated by the ELGs. For example, if your facility has recirculating systems and ponds, only the recirculating systems are regulated by the ELGs if they meet the production requirements. The requirements for your recirculating system will appear in your NPDES permit. The ELGs requirements will not apply to the ponds at your facility. However, you may need an NPDES permit if your ponds meet the definition of the cold water or warm water species category, where the ponds discharge at least 30 days per year (40 CFR part 122, Appendix C) or if your pond is part of a facility that has been designated a CAAP facility.

What systems are not regulated by the ELGs?

The ELGs regulations will not apply to you if you have systems other than flow-through, recirculating, or net pens, such as molluscan shellfish hatcheries or shrimp ponds. The regulations also do not apply to facilities whose combined annual production for their flow-through, recirculating, and net pen systems is less than 100,000 pounds. Systems not covered by the CAAP ELGs include:

- Closed pond systems (may be covered by NPDES if it discharges more than 30 days per year or it is designated as a CAAP facility by the Director)
- Molluscan shellfish (including nurseries)
- Crawfish production
- Alligator production
- Aquaria
- Net pens rearing native species released after a growing period of no longer than 4 months to supplement commercial and sport fisheries

What part of my CAAP is regulated?

The CAAP regulation applies to the production areas of your facility, including:

- Areas where you might grow, maintain, or contain aquatic animals (e.g., raceways, tanks, or net pens).
- Areas where you might store raw materials (e.g., feed silos and storage areas designated for feed or drugs).
- Areas where you might contain wastes (e.g., sedimentation basins, quiescent zones, and settling ponds).
- Source water and wastewater conveyance systems (e.g., tailraces and headraces).

When do I have to get an NPDES permit?

Your permit application deadline depends on whether your operation is an existing CAAP facility, a new discharger, or a new source. Each category has a different deadline for applying for an NPDES permit. Read the descriptions in Chapter 4 of this guide to determine when you must apply for an NPDES permit.

What records do I have to keep?

Your NPDES permit will require you to keep certain records to show that you are complying with the terms of the permit. You must keep all the records on-site at your operation for 5 years and you must provide them to the permitting authority upon request. Refer to Chapter 12 for additional information about record-keeping.

How does EPA treat proprietary/confidential information?

Disclosure of confidential business information (CBI) is restricted by statute. Pursuant to EPA regulations at 40 CFR 2.203 and 2.211, EPA treats all information for which a claim of confidentiality is made as confidential unless and until it makes a determination to the contrary under 40 CFR 2.205. Facilities that want to protect certain proprietary information included in their BMP plans should mark this information as CBI. Note that information from federal facilities and discharge information from all facilities cannot be claimed as CBI. Check with your permitting authority for process and eligibility information.

What monitoring do I have to perform under my NPDES permit?

The monitoring that your permitting authority will require as part of your permit will depend on the other conditions in your permit. If you are subject to ELGs, there are some associated monitoring requirements, which are routine inspections of production and wastewater treatment systems that are discussed in Chapters 5 and 11. In addition, if your permit includes water quality based numeric effluent limitations, you will be required to monitor to demonstrate compliance with those limitations. Your permitting authority may also require monitoring to characterize your discharge even when your permit does not include numeric effluent limitations. Look carefully at your permit, particularly the effluent limitations section, special conditions, and any specific monitoring requirements section, to determine what monitoring you have to perform.

How do I develop a BMP plan?

Your BMP plan should describe in detail how you will achieve the requirements of the CAAP ELGs. Chapter 8 of this guidance document describes what elements you should include in your BMP plan.

How do I certify my BMP plan?

Send a signed letter to your permitting authority, stating that you have developed a BMP plan. You will need to send a letter every time your permit is renewed. The BMP certification form should include your name and title, name of the facility, NPDES number, and date the BMP plan was developed. An example certification form that may be submitted to your permitting authority is available in Appendix F of this guidance document. Be sure to check with your permitting authority if you have any questions about certifying your BMP plan.

What do I do with my BMP plan, once it has been developed?

Once you have developed your BMP plan, keep a copy of the plan available in case your permitting authority requests a copy. You should also provide copies to employees so they can implement the BMPs.

Where can I find example forms to help me satisfy requirements of the CAAP ELGs (e.g., solids control) and that may be submitted to my permitting authority?

Appendices M through T contain examples forms for all requirements of the CAAP ELGs. The example forms are available from this document or online at http://www.epa.gov/guide/aquaculture.

How do I contact my permitting authority?

If you are unsure how to contact your permitting authority, Appendix A contains contact information by state.

How can I get copies of the rule or additional information?

You can get a copy of the final rule by contacting the Office of Water Resource center at 202-566-1729 or sending them an e-mail at center.water-resource@epa.gov. You can also write or call the National Service Center for Environmental Publications (NSCEP), U.S. EPA/NSCEP, P.O. Box 42419, Cincinnati, Ohio 45242-2419, (800) 490-9198, http://www.epa.gov/ncepihom. Finally, Appendix D of this document provides a copy of the federal regulations. You can get electronic copies of the preamble, rule, and major supporting documents at http://www.epa.gov/guide/aquaculture or in E-Docket at http://www.epa.gov/edocket. Once in the E-Docket system, select "search," then key in the docket identification number (OW-2002-0026).

Frequently Asked Questions for Your Permitting Authority

The following are frequently asked questions that you may want to ask your permitting authority:

- Will additional requirements (e.g., state, local, TMDL), in addition to EPA's regulations, be included in my NPDES permit? If so, what are they?
- What regulations for my state apply to my CAAP facility?
- How do I know if I am complying with all federal requirements that apply to point source discharges?
- How will I know if the Director designates my facility, which does not meet the requirements of the NPDES regulation or the CAAP ELGs, as a CAAP?
- If I discharge to a POTW and have an NPDES permit, what are the pretreatment requirements that I must meet?
- What forms do I need to fill out to apply for an NPDES permit or to renew my current NPDES permit? Where can I obtain these forms?
- Is my facility eligible for an existing general NPDES permit?
- When do I have to get an NPDES permit?
- What happens if I don't submit my application by the deadline?
- What do I have to do when I renew my permit?
- Am I considered a new source or new discharger? I don't understand the difference.
- What happens to my permit if I close my facility?
- What happens if I make significant changes (e.g., increase production level) at my operation?
- Will I need to perform monitoring?
- What information do I have to report to my permitting authority?

Appendix D1

40 CFR 122.24

later than 90 days after becoming defined as a CAFO; except that

(iii) If an operational change that makes the operation a CAFO would not have made it a CAFO prior to April 14, 2003, the operation has until April 13, 2006, or 90 days after becoming defined as a CAFO, whichever is later.

(4) *New sources.* New sources must seek to obtain coverage under a permit at least 180 days prior to the time that the CAFO commences operation.

(5) *Operations that are designated as CAFOs.* For operations designated as a CAFO in accordance with paragraph (c) of this section, the owner or operator must seek to obtain coverage under a permit no later than 90 days after receiving notice of the designation.

(6) *No potential to discharge.* Notwithstanding any other provision of this section, a CAFO that has received a ''no potential to discharge'' determination in accordance with paragraph (f) of this section is not required to seek coverage under an NPDES permit that would otherwise be required by this section. If circumstances materially change at a CAFO that has received a NPTD determination, such that the CAFO has a potential for a discharge, the CAFO has a duty to immediately notify the Director, and seek coverage under an NPDES permit within 30 days after the change in circumstances.

(h) *Duty to Maintain Permit Coverage.* No later than 180 days before the expiration of the permit, the permittee must submit an application to renew its permit, in accordance with § 122.21(g). However, the permittee need not continue to seek continued permit coverage or reapply for a permit if:

(1) The facility has ceased operation or is no longer a CAFO; and

(2) The permittee has demonstrated to the satisfaction of the Director that there is no remaining potential for a discharge of manure, litter or associated process wastewater that was generated while the operation was a CAFO, other than agricultural stormwater from land application areas.

[68 FR 7265, Feb. 12, 2003]

§ 122.24 Concentrated aquatic animal production facilities (applicable to State NPDES programs, see § 123.25).

(a) *Permit requirement.* Concentrated aquatic animal production facilities, as defined in this section, are point sources subject to the NPDES permit program.

(b) *Definition. Concentrated aquatic animal production facility* means a hatchery, fish farm, or other facility which meets the criteria in appendix C of this part, or which the Director designates under paragraph (c) of this section.

(c) *Case-by-case designation of concentrated aquatic animal production facilities.* (1) The Director may designate any warm or cold water aquatic animal production facility as a concentrated aquatic animal production facility upon determining that it is a significant contributor of pollution to waters of the United States. In making this designation the Director shall consider the following factors:

(i) The location and quality of the receiving waters of the United States;

(ii) The holding, feeding, and production capacities of the facility;

(iii) The quantity and nature of the pollutants reaching waters of the United States; and

(iv) Other relevant factors.

(2) A permit application shall not be required from a concentrated aquatic animal production facility designated under this paragraph until the Director has conducted on-site inspection of the facility and has determined that the facility should and could be regulated under the permit program.

[48 FR 14153, Apr. 1, 1983, as amended at 65 FR 30907, May 15, 2000]

§ 122.25 Aquaculture projects (applicable to State NPDES programs, see § 123.25).

(a) *Permit requirement.* Discharges into aquaculture projects, as defined in this section, are subject to the NPDES permit program through section 318 of CWA, and in accordance with 40 CFR part 125, subpart B.

(b) *Definitions.* (1) *Aquaculture project* means a defined managed water area which uses discharges of pollutants

Pt. 122, App. C

section 301(b)(2)(A), (C), (D), (E) and (F) of CWA, whether or not applicable effluent limitations guidelines have been promulgated. See §§ 122.44 and 122.46.

Industry Category

Adhesives and sealants
Aluminum forming
Auto and other laundries
Battery manufacturing
Coal mining
Coil coating
Copper forming
Electrical and electronic components
Electroplating
Explosives manufacturing
Foundries
Gum and wood chemicals
Inorganic chemicals manufacturing
Iron and steel manufacturing
Leather tanning and finishing
Mechanical products manufacturing
Nonferrous metals manufacturing
Ore mining
Organic chemicals manufacturing
Paint and ink formulation
Pesticides
Petroleum refining
Pharmaceutical preparations
Photographic equipment and supplies
Plastics processing
Plastic and synthetic materials manufacturing
Porcelain enameling
Printing and publishing
Pulp and paper mills
Rubber processing
Soap and detergent manufacturing
Steam electric power plants
Textile mills
Timber products processing

APPENDIX B TO PART 122 [RESERVED]

APPENDIX C TO PART 122—CRITERIA FOR DETERMINING A CONCENTRATED AQUATIC ANIMAL PRODUCTION FACILITY (§ 122.24)

A hatchery, fish farm, or other facility is a concentrated aquatic animal production facility for purposes of § 122.24 if it contains, grows, or holds aquatic animals in either of the following categories:

(a) Cold water fish species or other cold water aquatic animals in ponds, raceways, or other similar structures which discharge at least 30 days per year but does not include:

(1) Facilities which produce less than 9,090 harvest weight kilograms (approximately 20,000 pounds) of aquatic animals per year; and

(2) Facilities which feed less than 2,272 kilograms (approximately 5,000 pounds) of food during the calendar month of maximum feeding.

(b) Warm water fish species or other warm water aquatic animals in ponds, raceways, or other similar structures which discharge at least 30 days per year, but does not include:

(1) Closed ponds which discharge only during periods of excess runoff; or

(2) Facilities which produce less than 45,454 harvest weight kilograms (approximately 100,000 pounds) of aquatic animals per year. ''Cold water aquatic animals'' include, but are not limited to, the *Salmonidae* family of fish; e.g., trout and salmon. ''Warm water aquatic animals'' include, but are not limited to, the *Ameiuride, Centrarchidae* and *Cyprinidae* families of fish; e.g., respectively, catfish, sunfish and minnows.

Appendix D2

40 CFR 451

Monday,
August 23, 2004

Part II

Environmental Protection Agency

40 CFR Part 451
Effluent Limitations Guidelines and New Source Performance Standards for the Concentrated Aquatic Animal Production Point Source Category; Final Rule

ENVIRONMENTAL PROTECTION AGENCY

40 CFR Part 451

[OW–2002–0026; FRL–7783–6]

RIN 2040–AD55

Effluent Limitations Guidelines and New Source Performance Standards for the Concentrated Aquatic Animal Production Point Source Category

AGENCY: Environmental Protection Agency.

ACTION: Final rule.

SUMMARY: Today's final rule establishes Clean Water Act effluent limitations guidelines and new source performance standards for concentrated aquatic animal production facilities. The animals produced range from species produced for human consumption as food to species raised to stock streams for fishing. The animals are raised in a variety of production systems. The production of aquatic animals contributes pollutants such as suspended solids, biochemical oxygen demand, and nutrients to the aquatic environment. The regulation establishes technology-based narrative limitations and standards for wastewater discharges from new and existing concentrated aquatic animal production facilities that discharge directly to U.S. waters. EPA estimates that compliance with this regulation will affect 242 facilities. The rule is projected to reduce the discharge of total suspended solids by about 0.5 million pounds per year and reduce the discharge of biochemical oxygen demand (BOD) and nutrients by about 0.3 million pounds per year. The estimated annual cost for commercial facilities is $0.3 million. The estimated annual cost to Federal and State hatcheries is $1.1 million. EPA estimates that the annual monetized environmental benefits of the rule will be in the range of $66,000 to $99,000.

DATES: This regulation is effective September 22, 2004. For judicial review purposes, this final rule is promulgated as of 1 p.m. (Eastern time) on September 7, 2004 as provided at 40 CFR 23.2.

ADDRESSES: EPA has established a docket for this action under Docket ID No. OW–2002–0026. All documents in the docket are listed in the EDOCKET index at *http://www.epa.gov/edocket.* Although not listed in the index, some information is not publicly available, *i.e.,* confidential business information or other information whose disclosure is restricted by statute. Certain other material, such as copyrighted material, is not placed on the Internet and will be publicly available only in hard copy form. Publicly available docket materials are available either electronically in EDOCKET or in hard copy at the Water docket in the EPA Docket Center (EPA/DC) EPA West, Room B102, 1301 Constitution Ave., NW., Washington, DC. The EPA Docket Center Public Reading Room is open from 8:30 a.m. to 4:30 p.m., Monday through Friday, excluding legal holidays. The telephone number for the Public Reading Room is (202) 566–1744, and the telephone number for the Water Docket is (202) 566–2426.

FOR FURTHER INFORMATION CONTACT: For additional information contact Marta Jordan at (202) 566–1049.

SUPPLEMENTARY INFORMATION:

I. General Information

A. Does This Action Apply To Me?

Entities that directly discharge to waters of the U.S. potentially regulated by this action include:

Category	Examples of regulated entities and SIC Codes	Examples of regulated entities and NAICS codes
Facilities engaged in concentrated aquatic animal production, which may include the following sectors: Commercial (for profit) and Non-commercial (public) facilities.	0273—Animal Aquaculture. 0921—Fish Hatcheries and Preserves.	112511—Finfish Farming and Fish Hatcheries. 112519—Other Animal Aquaculture.

This table is not intended to be exhaustive, but rather provides a guide for readers regarding entities likely to be regulated by this action. This table lists the types of entities that EPA is now aware could potentially be regulated by this action. Other types of entities not listed in the table could also be regulated. To determine whether your facility is regulated by this action, you should carefully examine the applicability criteria listed at 40 CFR part 451 of today's rule. If you have questions regarding the applicability of this action to a particular entity, consult the person listed for information in the preceding **FOR FURTHER INFORMATION CONTACT** section.

B. How Can I Get Copies of This Document and Other Related Information?

1. Docket. EPA has established an official public docket for this action under Docket ID No. OW–2002–0026. The official public docket consists of the documents specifically referenced in this action, any public comments received, and other information related to this action. Although a part of the official docket, the public docket does not include Confidential Business Information (CBI) or other information whose disclosure is restricted by statute. The official public docket is the collection of materials that is available for public viewing at the Water Docket in the EPA Docket Center (EPA/DC), EPA West, Room B102, 1301 Constitution Ave., NW., Washington, DC. The EPA Docket Center Public Reading Room is open from 8:30 a.m. to 4:30 p.m., Monday through Friday, excluding legal holidays. The telephone number for the Public Reading Room is (202) 566–1744, and the telephone number for the Water Docket is (202) 566–2426. Every user is entitled to copy 266 pages per day before incurring a charge. The Docket may charge 15 cents a page for each page over the page limit plus an administrative fee of $25.00.

2. Electronic Access. You may access this Federal Register document electronically through the EPA Internet under the "Federal Register" listings at *http://www.epa.gov/fedrgstr/.*

An electronic version of the public docket is available through EPA's electronic public docket and comment system, EPA Dockets. You may use EPA Dockets at *http://www.epa.gov/edocket/* to view public comments, access the index listing of the contents of the official public docket, and to access those documents in the public docket that are available electronically. Once in the system, select "search," then key in the appropriate docket identification number. Although not all docket materials may be available electronically, you may still access any of the publicly available docket materials through the docket facility identified in section B.1.

C. What Other Information Is Available To Support This Final Rule?

The major documents supporting the final regulations are the following:

• "Technical Development Document for the Final Effluent Limitations Guidelines and New Source Performance Standards for the Concentrated Aquatic Animal Production Point Source Category" [EPA–821–R–04–012] referred to in the preamble as the Technical Development Document (TDD). The TDD presents the technical information that formed the basis for EPA's decisions in today's final rule. The TDD describes, among other things, the data collection activities, the wastewater treatment technology options considered by the Agency as the basis for effluent limitations guidelines and standards, the pollutants found in wastewaters from concentrated aquatic animal production facilities, the estimates of pollutant removals associated with certain pollutant control options, and the cost estimates related to reducing the pollutants with those technology options.

• "Economic and Environmental Benefit Analysis of the Final Effluent Limitations Guidelines and Standards for the Concentrated Aquatic Animal Production Point Source Category [EPA–821–R–04–013] referred to in this preamble as the Economic and Environmental Benefit Analysis or EEBA. This document presents the methodology used to assess economic impacts, environmental impacts and benefits of the final rule. The document also provides the results of the analyses conducted to estimate the projected impacts and benefits.

Major supporting documents are available in hard copy from the National Service Center for Environmental Publications (NSCEP), U.S. EPA/NSCEP, P.O. Box 42419, Cincinnati, Ohio, USA 45242–2419, (800) 490–9198, *www.epa.gov/ncepihom.* You can obtain electronic copies of this preamble and rule as well as major supporting documents at EPA Dockets at *www.epa.gov/edocket* and at *www.epa.gov/guide/aquaculture.*

D. What Process Governs Judicial Review for Today's Final Rule?

Under Section 509(b)(1) of the Clean Water Act (CWA), judicial review of today's effluent limitations guidelines and standards may be obtained by filing a petition for review in the United States Circuit Court of Appeals within 120 days from the date of promulgation of these guidelines and standards. For judicial review purposes, this final rule is promulgated as of 1 pm (Eastern time) on September 7, 2004 as provided at 40 CFR 23.2. Under section 509(b)(2) of the CWA, the requirements of this regulation may not be challenged later in civil or criminal proceedings brought by EPA to enforce these requirements.

E. What Are the Compliance Dates for Today's Final Rule?

Existing direct dischargers must comply with today's limitations based on the best practicable control technology currently available (BPT), the best conventional pollutant control technology (BCT), and the best available technology economically achievable (BAT) as soon as their National Pollutant Discharge Elimination System (NPDES) permits include such limitations. Generally, this occurs when existing permits are reissued. New direct discharging sources must obtain an NPDES permit for the discharge and comply with applicable new source performance standards (NSPS) on the date the new sources begin discharging. For purposes of NSPS, a source is a new source if it commences construction after September 22, 2004.

F. How Does EPA Protect Confidential Business Information (CBI)?

Certain information and data in the record supporting the final rule have been claimed as CBI and, therefore, EPA has not included these materials in the record that is available to the public in the Water Docket. Further, the Agency has withheld from disclosure some data not claimed as CBI because release of this information could indirectly reveal information claimed to be confidential. To support the rulemaking while preserving confidentiality claims, EPA is presenting in the public record certain information in aggregated form, masking facility identities, or using other strategies.

Table of Contents

II. Definitions, Acronyms, and Abbreviations Used in This Document

Act—The Clean Water Act.

Agency—U.S. Environmental Protection Agency.

AWQC—Ambient water quality criteria.

BAT—Best available technology economically achievable, as defined by section 304(b)(2)(B) of the Act.

BCT—Best conventional pollutant control technology, as defined by section 304(b)(4) of the Act.

BMP—Best management practice, as defined by section 304(e) of the Act.

BOD$_5$—Biochemical oxygen demand measured over a five day period.

BPJ—Best professional judgment.

BPT—Best practicable control technology currently available, as defined by section 304(b)(1) of the Act.

CAAP—Concentrated aquatic animal production.

CBI—Confidential business information.

CFR—Code of Federal Regulations.

CWA—33 U.S.C. §§ 1251 *et seq.*, as amended.

Conventional Pollutants—Constituents of wastewater as determined by Section 304(a)(4) of the CWA (and EPA regulations), *i.e.*, pollutants classified as biochemical oxygen demand, total suspended solids, oil and grease, fecal coliform, and pH.

Daily Discharge—The discharge of a pollutant measured during any calendar day or any 24-hour period that reasonably represents a calendar day.

Daily Maximum Limit—the highest allowable "daily discharge".

Direct Discharger—A facility that discharges or may discharge treated or untreated wastewaters into waters of the United States.

DMR—Discharge monitoring report; consists of the reports filed with the permitting authority by permitted dischargers to demonstrate compliance with permit limits.

DO—Dissolved oxygen.

ELG—Effluent limitations guidelines.

EQIP—Environmental Quality Incentives Program.

Existing source—For this rule, any facility from which there is or may be a discharge of pollutants, the construction of which is commenced before September 22, 2004.

Extralabel drug use—Actual use or intended use of a drug in an animal in a manner that is not in accordance with the approved label. The Federal Food, Drug, and Cosmetic Act allows veterinarians to prescribe extralabel uses of certain approved animal drugs and approved human drugs for animals under certain conditions. These conditions are spelled out in Food and Drug Administration regulations at 21 CFR Part 530. Among these requirements are that any extralabel use must be by or on the order of a veterinarian within the context of a veterinarian-client-patient relationship, must not result in violative residues in food-producing animals, and the use must be in conformance with the regulations. A list of drugs specifically prohibited from extralabel use appears at 21 CFR 530.41.

Facility—All contiguous property and equipment owned, operated, leased, or under the control of the same person or entity.

FAO—United Nations Food and Agriculture Organization.

FCR—Feed conversion ratio.

FDF—Fundamentally different factor.

FFDCA—Federal Food, Drug, and Cosmetic Act, 21 U.S.C. 301, *et seq.*, as amended.

FIFRA—Federal Insecticide, Fungicide and Rodenticide Act.

FR—Federal Register.

FTE—Full Time Equivalent Employee.

FWS—U.S. Fish and Wildlife Service.

INAD—Investigational new animal drug. A new animal drug (or animal feed containing a new animal drug) intended for testing or clinical investigational use in animals. Food and Drug Administration regulations limit the conditions under which such drugs may be used. 21 CFR 511, 514.

Indirect Discharger-A facility that discharges or may discharge wastewaters into a publicly-owned treatment works.

JSA/AETF—Joint Subcommittee on Aquaculture, Aquaculture Effluents Task Force.

lb(s)/yr—pound(s) per year.

NAICS—North American Industry Classification System. NAICS was developed jointly by the U.S., Canada, and Mexico to provide new comparability in statistics about business activity across North America.

NEPA—National Environmental Policy Act, 33 U.S.C. 4321, *et seq.*

NMFS—National Marine Fisheries Service.

NPDES Permit—A permit to discharge wastewater into waters of the United States issued under the National Pollutant Discharge Elimination System, authorized by Section 402 of the CWA.

NRCS—Natural Resources Conservation Service.

Nonconventional Pollutants—Pollutants that are neither conventional pollutants listed at 40 CFR 401 nor toxic pollutants listed at 40 CFR 401.15 and Part 423 Appendix A.

Non-water quality environmental impact—Deleterious aspects of control and treatment technologies applicable to point source category wastes, including, but not limited to air pollution, noise, radiation, sludge and solid waste generation, and energy used.

NRDC—Natural Resources Defense Council.

NSPS—New Source Performance Standards.

NTTAA—National Technology Transfer and Advancement Act, 15 U.S.C. 272 note.

OMB—Office of Management and Budget

Outfall—The mouth of conduit drains and other conduits from which a facility discharges effluent into receiving waters.

Pass through—a discharge that exits a POTW into waters of the United States in quantities or concentrations that alone or in conjunction with discharges from other sources, causes a violation of any requirement of the POTW's NPDES permit (including an increase in the magnitude or duration of a violation).

PCB—Polychlorinated biphenyls.

POC—Pollutants of Concern. Pollutants commonly found in aquatic animal production wastewaters. Generally, a chemical is considered as a POC if it was detected in untreated process wastewater at 5 times a baseline value in more than 10% of the samples.

Point Source—Any discernable, confined, and discrete conveyance from which

pollutants are or may be discharged. See CWA Section 502(14).

POTW(s)—Publicly owned treatment works. It is a treatment works as defined by Section 212 of the Clean Water Act that is owned by a State or municipality (as defined by Section 502(4) of the Clean Water Act). This definition includes any devices and systems used in the storage, treatment, recycling and reclamation of municipal sewage or industrial wastes of a liquid nature. It also includes sewers, pipes and other conveyances only if they convey wastewater to a POTW Treatment Plant. The term also means the municipality as defined in Section 502(4) of the Clean Water Act, which has jurisdiction over the Indirect Discharges to and the discharges from such a treatment works.

Priority Pollutant—One hundred twenty-six compounds that are a subset of the 65 toxic pollutants and classes of pollutants outlined pursuant to Section 307 of the CWA. 40 CFR Part 423, Appendix A.

PSES—Pretreatment standards for existing sources of indirect discharges, under Section 307(b) of the CWA, applicable to indirect dischargers that commenced construction prior to the effective date of a final rule.

PSNS—Pretreatment standards for new sources under Section 307(c) of the CWA.

QUAL2E—Enhanced Stream Water Quality Model.

RFA—Regulatory Flexibility Act, 5 U.S.C. 601, *et. seq.*

SBREFA—Small Business Regulatory Enforcement Fairness Act of 1996, Public Law 104–121.

SIC—Standard Industrial Classification, a numerical categorization system used by the U.S. Department of Commerce to catalogue economic activity. SIC codes refer to the products or groups of products that are produced or distributed, or to services that are provided, by an operating establishment. SIC codes are used to group establishments by the economic activities in which they are engaged. SIC codes often denote a facility's primary, secondary, tertiary, etc. economic activities.

TDD—Technical Development Document.

TSS—Total Suspended Solids.

U.S.C.—United States Code.

UMRA—Unfunded Mandates Reform Act of 1995, 2 U.S.C. 1501.

USDA—United States Department of Agriculture.

III. Under What Legal Authority Is This Final Rule Issued?

The U.S. Environmental Protection Agency is promulgating these regulations under the authority of Sections 301, 304, 306, 307, 308, 402, and 501 of the Clean Water Act, 33 U.S.C. 1311, 1314, 1316, 1318, 1342, and 1361.

IV. What Is the Statutory and Regulatory Background to This Rule?

A. Clean Water Act

Congress passed the Federal Water Pollution Control Act (1972), also known as the Clean Water Act (CWA), to "restore and maintain the chemical, physical, and biological integrity of the Nation's waters." (33 U.S.C. 1251(a)). The CWA establishes a comprehensive program for protecting our nation's waters. Among its core provisions, the CWA prohibits the discharge of pollutants from a point source to waters of the U.S. except as authorized by a National Pollutant Discharge Elimination System (NPDES) permit. The CWA also requires EPA to establish national technology-based effluent limitations guidelines and standards (effluent guidelines or ELG) for different categories of sources, such as industrial, commercial and public sources of waters. Effluent guidelines are implemented when incorporated into an NPDES permit. Effluent guidelines can include numeric and narrative limitations, including Best Management Practices, to control the discharge of pollutants from categories of point sources.

Congress recognized that regulating only those sources that discharge effluent directly into the nation's waters may not be sufficient to achieve the CWA's goals. Consequently, the CWA requires EPA to promulgate nationally applicable pretreatment standards that restrict pollutant discharges from facilities that discharge wastewater indirectly through sewers flowing to publicly-owned treatment works (POTWs). (See Section 307(b) and (c), 33 U.S.C. 1317(b) & (c)). National pretreatment standards are established only for those pollutants in wastewater from indirect dischargers that may pass through, interfere with, or are otherwise incompatible with POTW operations. Generally, pretreatment standards are designed to ensure that wastewaters from direct and indirect industrial dischargers are subject to similar levels of treatment. In addition, POTWs must develop local treatment limits applicable to their industrial indirect dischargers. Any POTWs required to develop a pretreatment program must develop local limits to implement the general and specific national pretreatment standards. Other POTWs must develop local limits to ensure compliance with their NPDES permit for pollutants that result in pass through or interference at the POTW. (See 40 CFR 403.5). Today's rule does not establish national pretreatment standards for this category, which contains very few indirect dischargers, because the indirect dischargers would be discharging mainly TSS and BOD, which the POTWs are designed to treat and which consequently, do not pass through. In addition, nutrients

discharged from CAAP facilities are in concentrations lower, in full flow discharges, and similar in off-line settling basin discharges, to nutrient concentrations in human wastes discharged to POTWs. The options EPA considered do not directly treat nutrients, but some nutrient removal is achieved incidentally through the control of TSS. EPA concluded POTWs would achieve removals of TSS and associated nutrients equivalent to those achievable by the options considered for this rulemaking and therefore there would be no pass through of pollutants in amounts needing regulation. In the event of pass through that causes a violation of a POTW's NPDES limit, the POTW must develop local limits for its users to ensure compliance with its permit.

Direct dischargers must comply with effluent limitations in NPDES permits. Technology-based effluent limitations in NPDES permits are derived from effluent limitations guidelines and new source performance standards promulgated by EPA, as well as occasionally from best professional judgment analyses. Effluent limitations are also derived from water quality standards. The effluent limitations guidelines and standards are established by regulation for categories of industrial dischargers and are based on the degree of control that can be achieved using various levels of pollution control technology.

EPA promulgates national effluent limitations guidelines and standards for major industrial categories generally for three classes of pollutants: (1) Conventional pollutants (*i.e.*, total suspended solids, oil and grease, biochemical oxygen demand, fecal coliform, and pH); (2) toxic pollutants (*e.g.*, toxic metals such as chromium, lead, nickel, and zinc; toxic organic pollutants such as benzene, benzo-*a*-pyrene, phenol, and naphthalene); and (3) Nonconventional pollutants (*e.g.*, ammonia-N, formaldehyde, and phosphorus). EPA considered the discharge of these classes of pollutants in the development of this rule. EPA is establishing BMP requirements for the control of conventional, toxic and Nonconventional pollutants. EPA considers development of four types of effluent limitations guidelines and standards for direct dischargers. The paragraphs below describe those pertinent to today's rule.

1. Best Practicable Control Technology Currently Available (BPT)—Section 304(b)(1) of the CWA

EPA may promulgate BPT effluent limits for conventional, toxic, and

nonconventional pollutants. For toxic pollutants, EPA typically regulates priority pollutants, which consist of a specified list of toxic pollutants. In specifying BPT, EPA looks at a number of factors. EPA first considers the cost of achieving effluent reductions in relation to the effluent reduction benefits. The Agency also considers the age of the equipment and facilities, the processes employed, engineering aspects of the control technologies, any required process changes, non-water quality environmental impacts (including energy requirements), and such other factors as the Administrator deems appropriate. (See CWA 304(b)(1)(B)). Traditionally, EPA establishes BPT effluent limitations based on the average of the best performance of facilities within the industry, grouped to reflect various ages, sizes, processes, or other common characteristics. Where existing performance is uniformly inadequate, EPA may establish limitations based on higher levels of control than currently in place in an industrial category, if the Agency determines that the technology is available in another category or subcategory and can be practically applied.

2. Best Conventional Pollutant Control Technology (BCT)—Section 304(b)(4) of the CWA

The 1977 amendments to the CWA required EPA to identify additional levels of effluent reduction for conventional pollutants associated with BCT technology for discharges from existing industrial point sources. In addition to other factors specified in Section 304(b)(4)(B), the CWA requires that EPA establish BCT limitations after consideration of a two-part "cost-reasonableness" test. EPA explained its methodology for the development of BCT limitations in July 1986 (51 FR 24974).

Section 304(a)(4) designates the following as conventional pollutants: Biochemical oxygen demand measured over five days (BOD_5), total suspended solids (TSS), fecal coliform, pH, and any additional pollutants defined by the Administrator as conventional. The Administrator designated oil and grease as an additional conventional pollutant on July 30, 1979 (44 FR 44501).

3. Best Available Technology Economically Achievable (BAT)—Section 304(b)(2) of the CWA

In general, BAT effluent limitations guidelines represent the best economically achievable performance of facilities in the industrial subcategory or category. The CWA establishes BAT as a principal national means of controlling the direct discharge of toxic and nonconventional pollutants. The factors considered in assessing BAT include the cost of achieving BAT effluent reductions, the age of equipment and facilities involved, the process employed, potential process changes, non-water quality environmental impacts including energy requirements, economic achievability, and such other factors as the Administrator deems appropriate. The Agency retains considerable discretion in assigning the weight to be accorded these factors. Generally, EPA determines economic achievability on the basis of total costs to the industry and the effect of compliance with BAT limitations on overall industry and subcategory financial conditions. As with BPT, where existing performance is uniformly inadequate, BAT may reflect a higher level of performance than is currently being achieved based on technology transferred from a different subcategory or category. BAT may be based upon process changes or internal controls, even when these technologies are not common industry practice.

4. New Source Performance Standards (NSPS)—Section 306 of the CWA

New Source Performance Standards reflect effluent reductions that are achievable based on the best available demonstrated control technology. New facilities have the opportunity to install the best and most efficient production processes and wastewater treatment technologies. As a result, NSPS should represent the most stringent controls attainable through the application of the best available demonstrated control technology for all pollutants (*i.e.*, conventional, nonconventional, and priority pollutants). In establishing NSPS, EPA is directed to take into consideration the cost of achieving the effluent reduction, any non-water quality environmental impacts, and energy requirements.

B. Section 304(m) Consent Decree

Section 304(m) of the CWA requires EPA every two years to publish a plan for reviewing and revising existing effluent limitations guidelines and standards and for promulgating new effluent guidelines. On January 2, 1990, EPA published an Effluent Guidelines Plan (see 55 FR 80) in which the Agency established schedules for developing new and revised effluent guidelines for several industry categories. Natural Resources Defense Council, Inc., and Public Citizen, Inc., challenged the Effluent Guidelines Plan in a suit filed in the U.S. District Court for the District

of Columbia, (*NRDC et al* v. *Leavitt*, Civ. No. 89–2980). On January 31, 1992, the court entered a consent decree which, among other things, established schedules for EPA to propose and take final action on effluent limitations guidelines and standards for several point source categories. The amended consent decree requires EPA to take final action on the Concentrated Aquatic Animal Production (CAAP) effluent guidelines by June 30, 2004.

C. Clean Water Act Requirements Applicable to CAAP Facilities

EPA's existing National Pollutant Discharge Elimination System (NPDES) regulations define when a hatchery, fish farm, or other facility is a concentrated aquatic animal production facility and, therefore, a point source subject to the NPDES permit program. See 40 CFR 122.24. In defining "concentrated aquatic animal production (CAAP) facility," the NPDES regulations distinguish between warmwater and coldwater species of fish and define a CAAP facility by, among other things, the size of the operation and frequency of discharge.

A facility is a CAAP facility if it meets the criteria in 40 CFR 122 appendix C or if it is designated as a CAAP facility by the NPDES program director on a case-by-case basis. The criteria described in appendix C are as follows. A hatchery, fish farm, or other facility is a concentrated aquatic animal production facility if it grows, contains, or holds aquatic animals in either of two categories: cold water species or warm water species. The cold water species category includes facilities where animals are produced in ponds, raceways, or other similar structures that discharge at least 30 days per year but does not include facilities that produce less than approximately 20,000 pounds per year or facilities that feed less than approximately 5,000 pounds during the calendar month of maximum feeding. The warm water species category includes facilities where animals are produced in ponds, raceways, or other similar structures that discharge at least 30 days per year, but does not include closed ponds that discharge only during periods of excess runoff or facilities that produce less than approximately 100,000 pounds per year. 40 CFR part 122, appendix C. Today's action does not revise the NPDES regulation that defines CAAP facilities.

Most facilities falling under the definition of CAAP are either flow-through, recirculating or net pen systems. These systems discharge continuously or discharge 30 days or

more per year as defined in 40 CFR part 122 and are subject to permitting depending on the production level at the facility. Most pond facilities do not require permits because ponds generally discharge fewer than 30 days per year and therefore generally are not CAAP facilities unless designated by the NPDES program director. The NPDES program director can designate a facility on a case-by-case basis if the director determines that the facility is a significant contributor of pollution to waters of the U.S.

V. How Was This Final Rule Developed?

This section describes the background to development of the proposal, the proposed rule, EPA's data collection effort, and changes to the proposal EPA considered based on new information and comments on the proposal.

A. September 2002 Proposed Rule

EPA started work on these effluent guidelines in January 2000. EPA relied on a federal interagency group known as the Joint Subcommittee on Aquaculture as a primary contact for information about the industry. The Joint Subcommittee on Aquaculture, authorized by the National Aquaculture Act of 1980, 94 Stat. 1198, 16 U.S.C. 2801, *et seq*, operates under the National Science and Technology Council of the Office of Science and Technology in the Office of the Science Advisor to the President. The National Aquaculture Act's purpose is to promote aquaculture in the United States to help meet its future food needs and contribute to solving world resource problems. The Act provides for the identification of regulatory constraints on the development of commercial aquaculture, and for development of a plan identifying specific steps the Federal Government can take to remove unnecessarily burdensome regulatory barriers to the initiation and operation of commercial aquaculture ventures. It also directs Federal agencies with functions or responsibilities that may affect aquaculture to perform such functions or responsibilities, to the maximum extent practicable, in a manner that is consistent with the purpose and policy of the Act. The Joint Subcommittee on Aquaculture established the Aquaculture Effluents Task Force (AETF) to work with EPA to provide information and expertise for the development of this rule. The AETF became an instrumental group providing input and comments to EPA. The AETF consists of members from various Federal agencies, State governments, industry, academia, and non-governmental (environmental) organizations.

EPA used the information provided by the AETF and conducted its own research for this rulemaking effort. EPA also relied on the 1998 Census of Aquaculture conducted by the Department of Agriculture (USDA) to provide information on the size and distribution of facilities in the industry. The Census also provided some basic information on the revenues and prices realized by aquatic animal producers. This information became a primary resource for describing the industry.

Because of limitations in the Census data, EPA conducted its own survey of the aquatic animal production industry. EPA adopted a two-phase approach to collecting data from aquatic animal producers. In the first phase, EPA distributed a "screener" survey. EPA designed this survey to collect very basic information from all known aquatic animal producers including public facilities regardless of size, ownership, or production system. EPA mailed the survey to approximately 6,000 potential aquatic animal producers in August 2001. The survey consisted of 11 questions asking for general facility information. EPA used the information collected to refine the profiles of the industry with respect to the production systems in use and the type of effluent controls in use. The screener survey, AETF information, and Census data became the primary sources for the proposed rule.

EPA based the limitations and standards for the proposed rule on the analysis of technologies to achieve effluent reductions using model aquatic animal production facilities. Each of these model facilities represented a different segment of the population corresponding to a particular production system type, size range (in terms of annual pounds of aquatic animals produced), and species produced.

EPA evaluated the economic impact of each regulatory option it considered for the proposed effluent limitations and new source performance standards based on the revenues and production cost information available from the USDA Census of Aquaculture along with EPA's own engineering cost estimates for the pollution control technologies being considered. After determining revenues and compliance costs for each model facility, EPA used a compliance cost-to-revenue ratio as a predictor of potential economic impacts for the different model facilities. EPA used this economic analysis in its evaluation of whether it should limit the

application of the national limitations and standards by size of production.

On September 12, 2002, EPA published the proposed rule (see 67 FR 57872). The proposed limitations and standards applied only to new and existing CAAP facilities that discharge directly to waters of the United States. EPA proposed requirements for three subcategories for this industry: flow-through, recirculating, and net pen systems. Flow-through and recirculating production systems are land-based. Net pens, by contrast, are located in open water.

EPA based the proposed requirements for the recirculating and flow-through subcategories on effluent control technologies that remove suspended solids from the animal production water prior to discharge. The technologies considered include quiescent zones, settling basins (including off-line settling basins, full flow settling basins, and polishing settling basins) and filtration technology. EPA proposed to establish limitations on the concentration of Total Suspended Solids (TSS) in the discharges from these facilities based on its preliminary assessment of the performance achieved by the various control technologies. In the case of recirculating systems, EPA based the proposed TSS limitations on solids polishing or secondary solids removal technology. For flow-through systems, EPA based the proposed TSS limitations on primary or secondary solids settling technologies depending on the production level of the facility (i.e., primary for 100,000–475,000 lbs/yr and secondary for >475,000 lbs/yr). In addition to numeric limits, EPA also proposed to require these facilities to implement operational measures so-called—Best Management Practices (BMPs)—to reduce the discharge of pollutants and develop a BMP plan to document these practices. Depending on the type and size of the facility, the plan would have required a facility to identify and implement practices that controlled, for example, the discharge of solids and ensured the proper storage and disposal of drugs and chemicals.

EPA based the proposed requirements for net pen facilities on requirements to reduce the amount of solids, mainly feed, being added directly into waters of the U.S. The proposal required net pen facilities to develop and implement BMPs to address the discharge of solids including the requirement to conduct active feed monitoring to minimize the amount of feed not eaten and thus discharged to the aquatic environment. Other proposed requirements included adoption of practices to ensure proper storage and disposal of drugs and

chemicals. In addition, EPA proposed that net pen facilities prevent the discharge of solid wastes such as feed bags, trash, net cleaning debris, and dead fish; chemicals used to clean the nets, boats or gear; and materials containing or treated with tributyltin compounds. Further requirements were designed to minimize the discharge of blood, viscera, fish carcasses or transport water containing blood associated with the transport or harvesting of fish.

B. December 2003 Notice of Data Availability

On December 29, 2003, EPA published a Notice of Data Availability (NODA) at 68 FR 75068. In the NODA, EPA summarized the data received since the proposed rule and described how the Agency might use the data for the final rule. The NODA also discussed the second phase of data collection, a detailed survey, which EPA conducted in 2002. The detailed survey was mailed to a stratified sample population of facilities identified from the screener survey. EPA received responses from 203 facilities. The surveyed population included a statistically representative sample of facilities that reported producing aquatic animals with flow-through, recirculating and net pen systems. EPA also surveyed a small number of facilities that would not have been subject to the proposed requirements. EPA's objective was to further verify the assumptions on which it had based its preliminary decision to exclude these facilities from the scope of the final rule.

The detailed data collected through this survey allowed EPA to revise the methods used for the proposed rule to estimate costs and economic impacts. EPA developed facility-specific costs and economic impact assessments for each surveyed facility based on the detailed information provided in the survey responses. The detailed information included production systems, annual production, and control practices and technologies in place at the facility.

The detailed responses to the second survey provided EPA with better information on the baseline level of control technologies and operational measures in use at CAAP facilities. Based on this understanding, EPA described two modified options in the NODA that EPA was considering for the final rule. These options reflected the same technologies and practices considered for the proposed regulation, but reconfigured the combinations of treatment technologies and practices into revised regulatory options.

EPA visited 17 additional sites and sampled at one facility in response to issues raised in the comments. The NODA discussed the post-proposal data including site visits and additional sampling. The results of EPA's analyses of the data were also presented in the NODA. EPA solicited comment on the new data and the conclusions being drawn from them.

C. Public Comments

EPA has prepared a "Comment Response Document" that includes the Agency's responses to comments submitted on the proposed rule and the notice of data availability. All of the public comments, including supporting documents, are available for public review in the administrative record for this final rule, filed under docket number OW–2002–0026.

The comment period on the proposed rule closed on January 27, 2003. EPA received approximately 300 comments, including form letters. EPA received comments from sources including the Joint Subcommittee on Aquaculture— Aquaculture Effluents Task Force (JSA/ AETF), industry trade associations, Federal and State agencies, environmental organizations, and private citizens. For the NODA, EPA received 20 comments between December 29, 2003 and February 12, 2004.

D. Public Outreach

As part of the development of the proposed rule and today's final rule, EPA has conducted outreach activities. EPA met with affected and interested stakeholders through site visits and sampling trips to obtain information on operating and waste management practices at CAAP facilities. EPA met numerous times with members of the JSA/AETF and conducted outreach with small businesses during the SBREFA process.

EPA conducted three public meetings to discuss the proposed rule during the public comment period for the proposed rule. EPA has participated in the industry's conferences to update participants on the progress and status of the rule. EPA also held several meetings with other federal agencies to discuss issues that potentially affect their mission, programs, or responsibilities.

Moreover, EPA maintains a website that posts information relating to the regulation. EPA provided supporting documents for the proposed rule on the site. The documents included the Technical Development Document, the Draft Guidance for Aquatic Animal Production Facilities to Assist in

Reducing the Discharge of Pollutants, and the Economic and Environmental Impact Analysis. These documents used to support the proposed rule and the final supporting documents are available at *www.epa.gov/guide/ aquaculture.*

VI. What Are Some of the Significant Changes in the Content of the Final Rule and the Methodology Used To Develop It?

This section describes some of the major changes that EPA made to the final rule from that it proposed. This section also describes differences in the methodology EPA used in evaluating its options for the final rule.

A. Subcategorization

The proposed regulation included limitations and standards for three subcategories: Flow-through systems, recirculating systems and net pens. The final rule establishes limitations and standards for the same systems but for only two subcategories: A flow-through and recirculating systems subcategory and a net pens subcategory. The recirculating and flow-through systems are combined into one subcategory instead of two separate subcategories.

As previously noted, flow-through and recirculating systems are both land based systems that typically discharge continuously, but can occasionally discontinue discharges for short periods of time. The principal distinguishing characteristic between these two systems is the degree to which water is reused prior to its discharge, with recirculating systems typically discharging lower volumes of wastewater. In the proposal, EPA distinguished recirculating systems from flow-through systems by describing a recirculating system as one that typically filters with biological or mechanically supported filtration and reuses the water in which the aquatic animals are raised. Net pen systems, by contrast, are located in open water and have distinctly different characteristics from either recirculating or flow-through systems.

EPA received a number of comments on the distinction between flow-through and recirculating systems described in the proposed rule. Because some flow-through systems also reuse their production water, commenters did not believe EPA had adequately distinguished recirculating systems from flow-through systems. Some commenters encouraged EPA to use hydraulic retention time as a basis for distinguishing between flow-through and recirculating systems. However, EPA's review of available data showed

that there is no clear dividing line between the hydraulic retention time in a system that was considered a recirculating system and one that was considered a flow-through system. EPA examined the aquatic animal production literature for alternatives for distinguishing recirculating systems and flow-through systems. Given the difficulty in distinguishing certain flow-through facilities from recirculating ones, EPA considered whether it should combine the two subcategories into one subcategory. EPA discussed this in the NODA and solicited comment on this option.

While some commenters opposed combining these two subcategories, EPA has decided to combine flow-through and recirculating systems for the purpose of establishing effluent limitations guidelines for the following reasons. First, as some commenters recognized, both flow-through and recirculating systems may reuse water and employ similar measures to maintain water quality including mechanical filtration. Second, the characteristic of wastewater discharged from facilities that are identified as recirculating systems that are similar to the wastewater from the off-line or solids treatment units at flow-through systems. Both waste streams are characterized by high levels of suspended solids, which can be effectively treated through properly designed and operated treatment systems employing either settling technology combined with effective feed management or a carefully controlled feed management system alone. Therefore, EPA decided that the same requirements should apply both to wastewater discharged from recirculating production systems and wastewater discharged from off-line solids treatment units at flow-through facilities. Moreover, EPA had based the proposed limits for both of these waste streams on the same data set. For the foregoing reasons, EPA has concluded that this change in the organization of the final rule does not substantively change the requirements.

Commenters also pointed to differences in BMPs employed at the different production systems. EPA recognizes that there are differences between recirculating systems and flow-through systems. EPA has concluded, however, that the control technology selected as the basis for the final narrative limitations will effectively remove pollutants from both systems to the same degree. Further, the BMP requirements in the final rule for this subcategory are flexible enough to accommodate differences in the specific practices appropriate for the two types of production systems. Finally, commenters were concerned that collapsing these two systems into one subcategory could be interpreted as indicating that EPA favors recirculating systems over flow-through systems and implying that flow-through systems should be modified to become recirculating systems. This certainly is not EPA's intention and the Agency is not suggesting that recirculating systems should replace existing flow-through systems or be given a preference in the construction of new systems. The primary reason to collapse these two systems into one subcategory is to eliminate redundancy in the CFR.

B. Regulated Pollutants

There are a number of pollutants associated with discharges from CAAP facilities. CAAP facilities can have high concentrations of suspended solids and nutrients, high BOD and low dissolved oxygen levels. Organic matter is discharged primarily from feces and uneaten feed. Metals, present in feed additives or from the deterioration of production equipment, may also be present in CAAP wastewater. Effluents with high levels of suspended solids, when discharged into receiving waters, can have a detrimental effect on the environment. Suspended solids can degrade aquatic ecosystems by increasing turbidity and reducing the depth to which sunlight can penetrate, thus reducing photosynthetic activity. Suspended particles can damage fish gills, increasing the risk of infection and disease. Nutrients are discharged mainly in the form of nitrate, ammonia and organic nitrogen. Ammonia causes two main problems in water. First, it is toxic to aquatic life. Second, it is easily converted to nitrate which may increase plant and algae growth.

Some substances, like drugs and pesticides, that may be present in the wastewater may be introduced directly as part of the aquatic animal production process. An important source of the pollutants potentially present in CAAP wastewater is, as the above discussion suggests, the feed used in aquatic animal production. Feed used at CAAP facilities contributes to pollutant discharges in a number of ways: by-product feces, ammonia excretions and, most directly, as uneaten feed (in dissolved and particulate forms). Moreover, the feed may be the vehicle for introducing other substances into the wastewater, like drugs. For example, medicated feed may introduce antibiotics into the wastewater.

In the proposed rule, EPA proposed to establish numeric limitations for only a single pollutant—total suspended solids (TSS)—while controlling the discharge of other pollutants through narrative requirements. Following proposal, EPA reevaluated the technological basis for the numerical limits for TSS and determined that it would be more appropriate to promulgate qualitative TSS limits, in the form of solids control BMP requirements, that could better respond to regional and site-specific conditions and accommodate existing state programs in cases where these appear to be working well (see Section VIII.B. for further discussion). EPA is thus not promulgating numerical limitations for TSS or other pollutants.

EPA is instead establishing narrative effluent limitations requiring implementation of effective operational measures to achieve reduced discharges of solids and other materials. For the final rule, as it did at proposal, EPA has also developed narrative limitations that will address a number of other pollutants potentially present in CAAP wastewater. These narrative limitations address spilled materials (drugs, pesticides and feed), fish carcasses, viscera and other waste, excess feed, feed bags, packaging material and netting.

EPA's decision to not establish national numeric limits for TSS will not restrict a permit writer's authority to impose site-specific permit numeric effluent limits on the discharge of TSS or other pollutants in appropriate circumstances. For example, a permit writer may establish water quality-based effluent limits for TSS (see 40 CFR 122.44(d) or regulate TSS (by establishing numeric limits) as a surrogate for the control of toxic pollutants (see 40 CFR 122.44(e)(2)(ii)) where site-specific circumstances warrant. The permit writer may also issue numeric limits in general permits applicable to classes of facilities. In fact, one of the bases for EPA's decision not to establish uniform national TSS limits is the recognition that a number of states, particularly those with significant numbers of CAAP facilities, already have general permits with numeric limits tailored to the specific production systems, species raised, and environmental conditions in the state, and these permits seem to be working well to minimize discharges of suspended solids (see DCN 63056). EPA believes there would be minimal environmental gain from requiring these states to redo their General Permits to conform to a set of uniform national concentration-based limits that in most cases would not produce significant changes in control technologies and practices at CAAP facilities.

In the final rule, EPA is also not establishing numeric limits for any drug or pesticide, but is requiring CAAP facilities to ensure proper storage of drugs, pesticides and feed to prevent spills and any resulting discharges of drugs and pesticides. EPA is also establishing a requirement to implement procedures for responding to spills of these materials to minimize their discharge from the facility. EPA's survey of this industry indicated that many CAAP facilities currently employ a number of different measures to prevent spills and have established in-place systems to address spills in the event they occur. EPA is thus establishing a requirement for all facilities to develop and implement BMPs that avoid inadvertent spills of drugs, pesticides, and feed and to implement procedures for properly containing, cleaning and disposing of any spilled materials to minimize their discharge from the facility. The effect of these requirements will be to promote increased care in the handling of these materials.

Some commenters suggested that EPA regulate certain other pollutants or substances that may be discharged from these production systems. For this rule, EPA evaluated control of some of these. For example, EPA evaluated the application of activated carbon treatment to remove compounds such as antibiotic active ingredients from wastewater prior to discharge. For the reasons discussed in Section IX.A, however, EPA is not basing any pollutant limitations on the application of this technology.

C. Treatment Options Considered

EPA evaluated three treatment options as the basis for BPT/BCT/BAT proposed limitations for the flow-through and recirculating subcategories and three options for the net pen subcategory. For flow-through and recirculating systems, EPA proposed a numeric limitation for TSS. For Option 1, the least stringent option, EPA considered TSS limitations based on primary settling as well as the use of BMPs to control the discharge of solids from the production system. The second treatment option (Option 2) considered by EPA for establishing TSS limitations was based on Option 1 technologies plus the addition of reporting requirements if INAD or extralabel drug use were used in the production systems, plus the implementation of BMPs to ensure proper storage, handling and disposal of drugs and chemicals and the prevention of escapes when non-native species are produced. EPA based limitations for the most stringent option (Option 3) on primary settling and the addition of secondary solids settling, in conjunction with BMPs, to control the discharge of solids from the production system. This option also included BMPs to control drugs, chemicals and non-native species and the reporting of drugs. For New Source Performance Standards (NSPS), EPA considered the same three options.

EPA evaluated three treatment options for the net pen subcategory. The least stringent option, Option 1, required feed management and operational BMPs for solids control. Option 2 consisted of the same practices and technology as Option 1 plus a BMP plan to address drugs, chemicals, pathogens, and non-native species and general reporting requirements for the use of certain drugs and chemicals. Option 3, the most stringent option, included the requirements of the first two options as well as active feed monitoring to control the supply of feed in the production units. Many existing facilities use active feed or real time monitoring to track the rate of feed consumption and detect uneaten feed passing through the nets. These systems may include the use of devices such as video cameras, digital scanning sonar detection, or upwellers, in addition to good husbandry and feed management practices. These systems and practices allow facilities to cease feeding the aquatic animals when a build-up of feed or over-feeding is observed. EPA considered the same treatment options for NSPS.

The NODA described two additional options that EPA was considering for flow-through and recirculating systems, but did not identify any new options for net pens. These two options contained the same treatment technologies and practices described in the three options considered for the proposed rule but in slightly different combinations.

The NODA Option A included primary solids treatment, a reporting requirement for the INAD and extralabel drug uses, and the implementation of BMPs to control drugs and chemicals. In addition to Option A requirements, Option B included secondary solids removal treatment or, alternatively, the implementation of BMPs for feed management, and solids handling to control the discharge of solids.

As previously explained, for flow-through or recirculating systems, today's final rule does not establish numeric limitations for total suspended solids (TSS) but does include narrative limitations requiring the solids control measures and operational practices described as part of Option B for BPT/BCT/BAT limitations and NSPS. These include requirements to minimize the

discharge of solids. It also requires facilities to develop and implement practices designed to prevent the discharge of spilled drugs and pesticides, inspection and maintenance protocols designed to prevent the discharge of pollutants as a result of structural failure, training of personnel, various recordkeeping requirements, and documentation of the implementation of these requirements in a BMP plan which is maintained on site and available to the permitting authority upon request.

For net pens, the final rule establishes non-numeric, narrative limitations that are similar to those adopted for flow-through and recirculating systems. Thus, the limitations require minimization of feed input, proper storage of drugs, pesticides and feed, routine inspection and maintenance of the production and wastewater treatment systems, training of personnel, and appropriate recordkeeping. Compliance with these requirements must be documented in a BMP plan which describes how the facility is minimizing solids discharges through feed management and how it is complying with prohibitions on the discharge of feed bags and other solid waste materials. Further, net pens must minimize the accumulation of uneaten feed beneath the pens through active feed monitoring and management strategies.

D. Reporting Requirements

EPA's proposed rule would have required permittees to report the use of INADs and extralabel use of both drugs and chemicals. In the final rule, EPA is modifying the proposed requirement, by deleting the reporting requirements for chemicals, including pesticides, and by further limiting the reporting requirement for drugs, as described below. EPA used the term "chemicals" in the proposed rule to refer to registered pesticides.

EPA's decision not to include pesticides in the final reporting requirements is based on the language in the Federal Insecticide, Fungicide and Rodenticide Act (FIFRA) and the regulations that implement the statute. FIFRA Section 5 authorizes EPA to allow field testing of pesticides under development through the issuance of Experimental Use Permits. Further, FIFRA Section 18 authorizes EPA to allow States to use a pesticide for an unregistered use for a limited time if EPA determines that emergency conditions exist. Under both of these provisions the applicant is required to submit information concerning the environmental risk associated with the

pesticide use as part of the application for the permit or exemption. Also in both cases the permittee or the State or Federal authority must report immediately to EPA any adverse effects from the use. Prior to issuing an emergency exemption, EPA is required to determine that the exemption will not cause unreasonable adverse effects on the environment (*see* 40 CFR 166.25(b)(1)(ii)) and that the pesticide is likely to be used in compliance with the requirements imposed under the exemption (*see* 40 CFR 166.25(b)(1)(iii)). EPA's regulation further specifies that the applicant for an emergency exemption must coordinate with other affected State or Federal agencies to which the requested exemption is likely to be of concern. The application must indicate that the coordination has occurred, and any comments provided by the other agencies must be submitted to EPA with the application (*see* 40 CFR 166.20(a)(8)).

In contrast, the FDA's regulations for Investigative New Animal Drugs (INADs) exempt INADs from the requirement to conduct an Environmental Assessment (*see* 21 CFR 25.20 and 25.33). As a policy matter, FDA encourages INAD sponsors to notify permitting authorities of the use of an INAD. There is, however, no requirement that the sponsors comply. Therefore, EPA considers the reporting of INADs in today's regulation necessary to ensure that permit writers are aware of the potential for discharge of the INAD and can take action as necessary in authorized circumstances.

EPA is providing an exception to the requirement to report INAD use. When an INAD has already been approved for use in another species or to treat another disease and is applied at a dosage that does not exceed the approved dosage, reporting is not required if it will be used under similar conditions. The requirement that the use be under similar conditions is intended to limit the exception to cases where the INAD use would not be expected to produce significantly different environmental impacts from the previously approved use. For example, use of a drug that had been previously approved for a freshwater application as an INAD in a marine setting would not be considered a similar condition of use, since marine ecosystems may have markedly different vulnerabilities than freshwater ecosystems. Similarly, the use of a drug approved to treat terrestrial animals as an INAD to treat aquatic animals would not be considered a similar condition of use. In contrast, the use of a drug to treat fish in a freshwater system that was previously approved for a different freshwater species would be considered use under similar conditions. EPA has concluded that when a drug is used under similar conditions it is unlikely that the environmental impacts would be different than those that were already considered in the prior approval of the drug.

CAAP facilities must also report the use of extralabel drugs. However, as with INADs, reporting is not required if the extralabel use does not exceed the approved dosage and is used under similar conditions. EPA anticipates that most extralabel drug use will not require reporting, but wants to ensure that permitting authorities are aware of situations in which a higher dose of a drug is used or the drug is used under significantly different conditions from the approved use. It is also possible that drugs approved for terrestrial animals could be used to treat aquatic animals as extralabel use drugs.

For the final rule, the timing and content of reporting requirements related to the use of INADs and extralabel drugs are similar to the proposed requirements. EPA requires both oral and written reporting. The final rule has an added requirement that the CAAP facility report the method of drug application in both the oral report and the written report. EPA has concluded that both oral and written reports are reasonable requirements because the oral report lets the permitting authority know of the drug use sooner than the written report, thus facilitating site-specific action if warranted. The written report provides confirmation of the use of the drug and more complete information for future data analysis and control measures. Today's regulation also adds a requirement that CAAP facilities notify the permitting authority in writing within seven days after signing up to participate in INAD testing. Advance notice prior to the use of the INAD allows the permitting authority to determine whether additional controls on the discharge of the INAD during its use may be warranted.

Finally, today's regulation includes a requirement to report any spill of drugs, pesticides or feed that results in a discharge to waters of the U.S. Facilities are expected to implement proper storage for these products and implement procedures for the containing, cleaning and disposing of spilled material. If the spilled material enters the production system or wastewater treatment system it can be assumed that the material will reach waters of the U.S. EPA considers reporting of these events necessary to alert the permitting authority to

potential impacts in the receiving stream. Facilities are expected to make an oral report to the permitting authority within 24 hours of the spill's occurrence followed by a written report within 7 days. The report shall include the identity of the material spilled and an estimated amount.

EPA has concluded that today's reporting requirements are appropriate because they make it easier for the permitting authority to evaluate what additional control measures on INADs and extralabel drug use may be necessary to prevent or minimize harm to waters of the U.S. and to respond more effectively to any unanticipated environmental impacts that may occur. Because neither of these classes of drugs has undergone an environmental assessment for the use being made of them, EPA is ensuring that the permitting authority is aware of their use and if warranted can take site specific action.

Today's reporting requirements are authorized under several sections of the CWA. Section 308 of the CWA authorizes EPA to require point sources to make such reports and "provide such other information as [the Administrator] may reasonably require." 33 U.S.C. 1318(a)(A). Section 402(a) of the Act authorizes EPA to impose permit conditions as to "data and information collection, reporting and such other requirements as [the Administrator] deems appropriate." 33 U.S.C. 1342(a)(2). It is well established that these provisions justify EPA's establishing a range of information disclosure requirements. Thus, for example, the United States Court of Appeals for the District of Columbia Circuit concluded that the Agency's data gathering authority was not limited to information on toxic pollutants already identified by the Agency in a permittee's discharge. EPA regulations required permit applications to include information on toxic pollutants that an applicant used or manufactured as an intermediate or final product or byproduct. In the court's view, EPA could reasonably determine that it could not regulate effectively without information on such pollutants because they could end up present in the permittee's discharge. *Natural Resources Defense Council, Inc.* v. *U.S. Environmental Protection Agency,* 822 F.2d 104, 119 (DC Cir. 1987). The same is true for certain INADs and extralabel drug use that may end up as pollutants discharged to waters of the U.S.

Under the proposed rule, the operators of facilities subject to the rule were to certify that they had developed a BMP plan that met the requirements

in the regulation. EPA continues to view BMPs as effective tools to control the discharge of pollutants from CAAP facilities and is establishing narrative requirements based on the use of BMPs as the basis of today's regulation. EPA has also retained the requirement for a BMP plan. The BMP plan is a tool in which the facility must describe the operational measures it will use to meet the non-numeric effluent limitations in the regulation. Upon incorporation of today's requirements into an NPDES permit, the CAAP facility owner or operator will be expected to develop site-specific operational measures that satisfy the requirements. The final rule requires CAAP facilities to develop a BMP plan that describes how the CAAP facility will comply with the narrative requirements and that is maintained at the CAAP facility. The CAAP facility owner or operator must certify in writing to the permitting authority that the plan has been developed. In EPA's view, a BMP plan, as a practical matter, can assist facilities in achieving compliance with the non-numeric limitations. It can also assist regulatory authorities in verifying compliance with the requirements and modifying specific permit conditions where warranted. As explained earlier in this section, EPA has concluded Section 308 clearly authorizes it to require this information. Of course, irrespective of the content of the plan, a facility must still comply with the narrative limitations.

In conjunction with the requirement to inspect and provide regular maintenance of CAAP production and treatment systems to prevent structural damage, EPA is including a reporting requirement associated with failure of the CAAP containment structure and any resulting discharges. EPA is requiring CAAP facilities to report any failure of or damage to the structural integrity of the containment system that results in a material discharge of pollutants to waters of the U.S. For net pen systems, for example, failures might include physical damage to the predator control nets or the nets containing the aquatic animals, that may result in a discharge of the contents of the nets. Physical damage might include abrasion, cutting or tearing of the nets and breakdown of the netting due to rot or ultra violet exposure. For flow-through and recirculating systems, a failure might include the collapse of, or damage to, a rearing unit or wastewater treatment structure; damage to pipes, valves, and other plumbing fixtures; and damage or malfunction to screens or physical barriers in the system, which would prevent the unit from containing water, sediment, and the aquatic animals. The permitting authority may further specify in the permit what constitutes a material discharge of pollutants that would trigger the reporting requirements. The permittee must report the failure of the containment system within 24 hours of discovery of the failure. The permittee must notify the permitting authority orally and describe the cause of the failure in the containment system and identify materials that were discharged as a result of this failure. Further, the facility must provide a written report within seven days of discovery of the failure documenting the cause, the estimated time elapsed until the failure was repaired, an estimate of the material released as a result of the failure, and steps being taken to prevent a reoccurrence.

E. Costs

At proposal, EPA used a model facility approach to estimate the cost of installing or upgrading wastewater treatment to achieve the proposed requirements. As described in the preamble to the proposed regulation (67 FR 57872), EPA developed 21 model facilities (based on the USDA's Census of Aquaculture and EPA's screener survey) characterized by different combinations of production systems, size categories, species and ownership types. EPA developed regulatory technology options based on screener survey responses, site visits, industry and other stakeholder input, and existing permit requirements.

EPA estimated the cost for each option component for each model facility. We then calculated costs for each regulatory option at each model facility based on model facility characteristics and the costs of the option's technologies or practices corresponding to the option.

EPA estimated frequency factors for treatment technologies and existing BMPs based on screener survey responses, site visits, and sampling visits. Baseline frequency factors represented the portion of the facilities represented by a particular model facility that would not incur costs to comply with the proposed requirements because they were already using the technology or practice. EPA adjusted the component cost for each model facility to account for those facilities that already have the component in-place. Subsequently, EPA derived national estimates of costs by aggregating the component costs applicable to each model facility across all model facilities.

EPA's detailed surveys captured information on the treatment in-place at the facility and other site-specific information (such as labor rates). EPA obtained additional cost information from data supplied from public comments and site visits. With the new data, EPA revised the method to estimate compliance costs. Instead of a model facility approach, EPA used a facility-level cost analysis based on the available facility-specific data contained in the detailed survey responses. We applied statistically-derived survey weights instead of the frequency factors used at proposal to estimate costs to the CAAP industry as a whole.

For proposal, EPA used national averages for many of the cost elements, such as labor rates and land costs. In its analysis for the final regulation, EPA used facility specific cost information, such as labor rates, to determine the costs associated with implementing the regulatory options. When facility specific rates were not available, EPA used national averages for similar ownership types of facilities (*i.e.*, non-commercial and commercial ownership) to determine managerial and staff labor rates. EPA revised estimates for all labor costs using the employee and wage information supplied in the detailed surveys. For those facilities indicating they use unpaid labor for part of the facility operation, we used wages for similar categories (*i.e.*, managerial or staff) supplied by that facility to estimate costs associated with implementing the regulatory options.

Comments also suggested that EPA's assumed land costs were too low at proposal; EPA assumed national average land values for agricultural land. EPA revised its estimates for land costs when determining the opportunity costs of using land at a facility if structural improvements were evaluated that required use of facility land that was not currently in use by the CAAP operation's infrastructure (*e.g.*, occupied by tanks, raceways, buildings, settling basins, *etc.*). When evaluating the cost of land for the revised analyses, EPA used land costs of $5,000/acre, which is twice the median value for land associated with aquaculture facilities surveyed in the U.S. (*see* DCN 63066). EPA used this conservative estimate because the only facilities that required structural improvements in the options evaluated were non-commercial facilities, for which land value estimates were not available.

EPA considered several technology-based options to determine the technical and economic feasibility of requiring numeric TSS limits for in-scope CAAP facilities. EPA's analysis of the detailed survey revealed that over 90% of the flow-through and recirculating system facilities currently had at least primary settling technologies in-place. EPA performed a cost analysis for the facilities without primary settling using the facility-specific configuration information provided in the detailed survey. EPA also evaluated facilities with primary settling in-place by comparing actual (*i.e.*, DMR data) or estimated TSS effluent concentrations to the proposed limits. For those facilities not meeting the proposed TSS limits, EPA also evaluated the implementation of additional solids controls, including secondary solids polishing and feed management.

For facilities with no solids control equipment, we estimated the costs for primary solids control. EPA evaluated each facility to identify the configuration of the existing treatment units and what upgrades would be required.

EPA also used industry cost information provided through public comment and the detailed survey to estimate costs for design and installation of primary settling equipment for effective settling of suspended solids. For example, we used the facility-level data included in the detailed survey responses to place and size the off-line settling basins on the facility site.

EPA classified each facility's wastewater treatment system based on the description provided in its survey response and available monitoring data, including DMR data. We assumed that treatment technologies indicated by a facility on the detailed survey are properly sized, installed, and maintained. EPA estimated facility-specific costs for each of the responding direct dischargers and used these estimates as the basis for national estimates. Because the survey did not collect information about many specific parameters used in individual facilities' production processes and treatment systems, EPA supplemented the facility-specific information with typical specifications or parameters from literature, survey results, and industry comments. For example, EPA assumed that facilities have pipes of typical sizes for their operations.

As a consequence of such assumptions, a particular facility might need a different engineering configuration from those modeled if it installed equipment that varies from the equipment or specifications we used to estimate costs. EPA nonetheless considers that costs for these facilities are generally accurate and representative, especially industry-wide. EPA applied typical specifications and parameters representative of the industry to a range of processes and treatment systems. We contacted facilities to get site-specific configuration information where possible.

In revising cost estimates, EPA paid particular attention to:

1. Size of tanks, raceways, and culture units;
2. Labor rates;
3. Treatment components in place;
4. BMPs and plans in place;
5. Daily operations at the facility.

Site visits and analysis of the detailed surveys indicated that raceways and quiescent zones are cleaned as necessary to maintain system process water quality.

In evaluating facilities for the need to use additional solids controls, EPA first checked for evidence of a good feed management program. If the facility reported they practice feed management, EPA looked for evidence of solids management and good operation of the physical plant, including regular cleaning and maintenance of feed equipment and solids collection devices (*e.g.*, quiescent zones, sedimentation basins, screens, *etc.*). To evaluate the effectiveness of a facility's solids control practices, we calculated feed conversion ratios (FCRs) using pounds of feed per pound of live product (as reported in the detailed survey) and considered existing solids control equipment. We assumed facilities lacking evidence of good feed management or solids control programs would incur additional costs to improve or establish them.

EPA estimated FCRs from data in the detailed survey and follow-up with some facilities and compared FCRs for groups of facilities (*i.e.*, combinations of ownership, species and production system types such as commercial trout flow-through facilities or government salmon flow-through facilities). We found a wide range of FCRs (reported by facilities in their detailed surveys, which were validated by call backs to the facility) among apparently similar facilities within ownership-species-production system groupings.

For example, we had good data for 24 of 60 government trout producers using flow-through systems. They reported a range of FCRs of 0.79 to 1.80 with a median FCR of 1.30. If an individual facility's reported FCR was significantly greater than the median, EPA further evaluated the facility to ascertain the reason for the higher FCR. Facilities that produce larger fish, such as broodstock, might have higher FCRs because the larger fish produce less flesh per unit of food. Facilities with fluctuating water temperatures could also be less efficient

than facilities with constant water temperatures. We did not apply costs for solids control BMPs for facilities with reasonable explanations for the higher FCRs. We evaluated facilities that did not report FCRs or provide enough data for an estimate by using a randomly selected FCR, which is described in Chapter 10 of the Technical Development Document (DCN 63009).

For those facilities that required additional solids controls, EPA evaluated both feed management and the installation of secondary solids polishing technologies. EPA received comments on the use of microscreen filters and EPA agrees with concerns raised in comments that the cost associated with enclosing the filter in a heated structure would be prohibitive. EPA found that the effective operation of microscreen filters requires that they be enclosed in heated buildings to prevent freezing when located in cold climates. EPA's revised estimates of costs for secondary solids polishing are not based on the application of microscreen filters unless the detailed survey response indicated that such a structure existed at the site. When the detailed survey did not indicate a structure at the site, EPA estimated costs for a second stage settling structure rather than a microscreen filter. Based on data from two of EPA's sampling episodes at CAAP facilities, this technology will achieve the proposed limits for TSS.

We also considered the use of activated carbon filtration to treat effluent containing drug or pesticide active ingredients from wastewater, but rejected controls for these materials. Research indicates that this technology is effective at treating these compounds, and at least one aquatic animal production facility installed this technology for water quality reasons. EPA estimated the costs for activated carbon treatment as a stand-alone technology. We estimated costs on a site-specific basis for facilities which reported using drugs and then added these costs for the different regulatory options considered to assess the economic achievability of this technology. A detailed discussion of how EPA estimated costs is available from the public record (DCN 62451). EPA considers these costs to be economically unachievable or not affordable on a national scale. However, EPA is aware of at least one facility currently using this technology, and notes that it is an effective technology for removing drug compounds from wastewater.

EPA estimated the costs to develop and implement escape management

practices at facilities where (1) the cultured species was not commonly produced or regarded as native in the State, (2) the facility was a direct discharger, and (3) the species was expected to survive if released. (In contrast, producers of a warm water species in a cold climate, such as tilapia producers in Minnesota or Idaho, would not incur costs for this practice.) Costs for escape prevention include staff time for production unit and discharge point inspections and maintenance of escape prevention devices. We applied these costs to facilities that installed equipment conforming with State requirements for facilities producing non-native species (identified by the State). Management time includes quarterly production unit and discharge point inspections, eight hours a year to review applicable State and Federal regulations, and quarterly staff consultations.

F. Economic Impacts

There are a number of changes made to the costing and economic impact methods used for the final rule. EPA used data from the detailed survey to project economic impacts for the final rule, in contrast to the screener data and frequency factors used for the proposed rule. For existing commercial operations, EPA assessed the number of business closures among regulated enterprises, facilities, and companies by applying market forecasts and using a closure methodology that compares projected earnings with and without incremental compliance costs for the period 2005 to 2015. Other additional analyses include an analysis of moderate impacts by comparing annual compliance costs to sales, an evaluation of financial health using a modified U.S. Department of Agriculture's four-category (2 × 2) matrix approach, and an assessment of possible impacts on borrowing capacity. For new commercial operations, EPA evaluates whether the regulatory costs will result in a barrier to entry among new businesses. For noncommercial operations, EPA evaluated impacts using a budget test that compares incurred compliance costs to facility operating budgets. Additional analyses investigate whether a facility could recoup increased compliance costs through user fees and estimated the associated increase.

For today's final regulation, EPA modified its forecasting models to include certain data for recent years that became available after the Agency published its NODA (*see* 68 FR 75068–75105). This and other details about how EPA developed its economic impact methodologies is presented in this preamble and in the Economic and Environmental Benefit Analysis of the Final Effluent Limitations Guidelines and Standards for the Concentrated Aquatic Animal Production Industry ("Economic and Environmental Benefit Analysis"), available in the rulemaking record.

G. Loadings

To estimate the baseline discharge loadings and load reductions for the proposed rule, EPA used the same model facility approach as used to estimate the compliance costs. Briefly, EPA first estimated pollutant loadings for untreated wastewater based on several factors for each model facility. As previously noted, feed used at CAAP facilities contributes to pollutant discharges in three ways: By-product feces, dissolved ammonia excretions, and uneaten feed (in dissolved and particulate forms). These byproducts of feed contribute to the pollutant load in the untreated culture water. EPA then used typical efficiency rates of removing specific pollutants from water to estimate load reductions for the treatment options and BMPs. EPA estimated frequency factors for treatment technologies and existing BMPs based on screener survey responses, site visits, and sampling visits. The occurrence frequency of practices or technologies was used to estimate the portion of the operations that would incur costs. Using the same frequency factors for technologies in place that were used to estimate costs, EPA estimated the baseline pollutant loads discharged, then calculated load reductions for the options.

As described in the NODA, EPA revised the loadings approach to incorporate facility-level information using data primarily from the detailed surveys. EPA also incorporated information included in comments concerning appropriate feed conversion ratios (FCRs).

EPA based its estimates of pollutant loads on the reported feed inputs included in the detailed surveys. EPA used the annual feed input and feed-to-pollutant conversion factors described in the TDD and DCN 63026 to calculate raw pollutant loads. EPA then analyzed each facility's detailed survey response to determine the treatment-in-place at the facility. Using published literature values to determine the pollutant removal efficiencies for the types of wastewater treatment systems used at CAAP facilities, EPA calculated a baseline pollutant load discharged from each surveyed facility. EPA used these pollutant removal efficiencies and raw

pollutant loads to estimate the baseline loads. EPA validated the baseline load estimates with effluent monitoring data (DCN 63061).

For today's regulation, EPA evaluated secondary solids removal technologies and feed management. EPA assessed whether improved feed management in addition to primary solids settling might be as effective at reducing solids in the effluent as secondary settling. EPA found that feed management was the lower cost option compared to secondary solids removal technology. (As discussed in more detail below at VIII.B., EPA has now concluded that a rigorous feed management program alone will achieve significant reductions in solids at CAAP facilities.)

Pollutant removals associated with feed management result from more efficient feed use and less wasted feed. For its evaluation, EPA used feed conversion rates as a surrogate for estimating potential load reductions resulting from feed management activities. Note, EPA used FCR values as a means to estimate potential load reductions, not as a target to set absolute FCR limits for a facility or industry segment.

Based on the information in the detailed surveys, EPA calculated FCRs for 69 flow-through and recirculating system facilities. EPA validated the feeding, production and estimated FCRs by contacting each facility. For those facilities that were not able to supply accurate feed and/or production information, to enable EPA to estimate a FCR, EPA randomly assigned a FCR.

EPA attempted to capture and account for as much of the variation as possible when analyzing FCRs and in the random assignment process. For example, the production system, species, and system ownership (which are all known from the detailed surveys) were expected to influence feeding practices, so facilities were grouped according to these parameters. EPA included ownership as a grouping variable to account for some of the variation in production goals. Most commercial facilities that were evaluated are producing food-sized fish and generally are trying to maintain constant production levels at the facility; commercial facilities would tend to target maximum weight gain over a low FCR in determining their optimal feeding strategy. Non-commercial facilities are generally government facilities that are producing for stock enhancement purposes. Production goals are driven by the desire to produce a target size (length and weight) at a certain time of year for release. Non-commercial facility feeding

goals may not place as great an emphasis on maximum growth. However, EPA expects that all facilities, regardless of production goals, can achieve substantial reductions in pollutant discharges over uncontrolled levels by designing and implementing an optimal feed input management strategy, including appropriate recordkeeping and documentation of FCRs.

The process for the random assignment of FCRs to facilities with incomplete information included:

• EPA grouped facilities by ownership, species, and production

• FCRs were estimated for each facility with sufficient data within a group

• The distributions of grouped data were examined for possible outliers, which were defined as FCRs less than 0.75 or greater than 3.0. When extreme values were found and validated, they were removed from the grouping. Although these extremes may be possible and a function of production goals, water temperature, etc., EPA was not able to validate and model all of the factors contributing to the extreme FCR rates. Facilities excluded because of extreme values were not assigned a random FCR, but were found to have a documented reason for the extreme value. For example, one facility produced broodstock for stock enhancement purposes. Some extreme values were updated based on validating information from the facility, and the updates were found to be within the range used for analysis.

• After removing outliers, the first and third quartiles were calculated for each grouping. The first quartile of a group of values is the value such that 25% of the values fall at or below this value. The third quartile of a group of values is the value such that 75% of the values fall at or below this value.

• For each grouping, the target FCR was assumed to be the first quartile value.

• For the facilities with no FCR information, a random FCR between the first and third quartiles was assigned.

• To account for variation in FCRs based on factors such as water temperature, EPA only costed additional feed management practices at a facility when the reported or randomly assigned FCR was within the upper 25% of the inter-quartile range. This was considered to be an indication of potential improvement in feed management.

• For some combinations of ownership, species, and production, there was not sufficient data to do the quartile analysis. In these cases, data from a similar grouping of ownership, species, and production was used.

If a facility's FCR was in the upper 25% of the inter-quartile range or did not currently have secondary settling technologies in place, EPA assumed the facility would need to improve feed management practices. The improvement in feed management practices would result in increased costs due to increased observations and recordkeeping and in pollutant load reductions resulting from less wasted feed.

The approach for estimating the loadings for the final rule has not changed significantly from the approach taken in the NODA. In estimating the loadings and removals for the final rule, EPA considered incidental removals or removals gained from the control of solids through narrative limitations. As part of the loadings analysis, EPA considered incidental removals of metals, PCBs and one drug, oxytetracycline.

Metals may be present in CAAP effluents from a variety of sources. Some metals are present in feed (as federally approved feed additives), occur in sanitation products, or may result from deterioration of CAAP machinery and equipment. EPA has observed that many of the treatment measures used in the CAAP industry provide substantial reductions of most metals. The metals present are generally readily adsorbed to solids and can be adequately controlled by controlling solids.

Most of the metals appear to be originating from the feed ingredients. Trace amounts of metals at federally approved concentrations are added to feed in the form of mineral packs to ensure that the essential dietary nutrients are provided for the cultured aquatic animals. Examples of metals added as feed supplements include copper, zinc, manganese, and iron (Snowden, 2003).

EPA estimated metals load reductions from facilities that are subject to the final rule (*see* DCN 63011). The metals for which load reductions are analyzed are those which were present above the detection levels in the wastewater samples collected from CAAP facilities during EPA's sampling for this rulemaking. EPA used the net concentrations of the metal in the wastewater to estimate these loads. EPA estimated these load reductions as a function of TSS loads using data obtained from the four sampling episodes. For this analysis, EPA first assumed that non-detected samples had the concentration of half the detection limit. From the sampling data, EPA calculated net TSS and metals

concentrations at different points in the facilities. EPA then calculated metal to TSS ratios (in mg of metal per kg of TSS) based on the calculated net concentrations. EPA removed negative and zero ratios from the samples. Finally, basic sample distribution statistics were calculated to derive the relationship between TSS and each metal.

EPA calculated estimated load reductions of PCBs from regulated facilities as a percentage of TSS load reductions. Since the main source of PCBs at CAAP facilities is through fish feed, a conversion factor was calculated to estimate the amount of PCBs discharged per pound of TSS. EPA assumed that 90% of the feed was eaten, and that 90% of the feed eaten would be assimilated by the fish. By combining the amount of food materials excreted by fish (10% of feed consumed) with the 10% of food uneaten, EPA was able to partition the PCBs among fish flesh and aqueous and solid fractions. Due to a lack of sampling data, EPA used a maximum level of 2μg/g, the FDA limit on PCB concentrations in fish feed, to estimate the maximum amount of PCBs that could possibly be in the TSS. This maximum possible discharge load in the TSS was estimated to be 21% of the PCBs in the feed. EPA considers this estimate to provide an upper bound on the amount of PCBs discharged from CAAP facilities, and the amount potentially removed by the rule. Even so, the estimates are quite low (0.52 pounds of PCBs discharged in the baseline). CAAP facilities are not a significant source of PCB discharges to waters of the U.S. (*see* DCN 63011).

EPA estimated the pollutant load of oxytetracycline discharged from in-scope CAAP facilities using data from EPA's detailed survey of the CAAP Industry. EPA first determined facility specific amounts of oxytetracycline used by each CAAP facility. For those facilities that reported using medicated feed containing oxytetracycline, EPA evaluated their responses to the detailed survey to determine the amount, by weight, of medicated feed containing oxytetracycline and the concentration of the drug in the feed. EPA then estimated the amount of oxytetracycline that was reduced at facilities in which feed management practices were applied in the cost and loadings analyses. The facility level estimates were then multiplied by the appropriate weighting factors and summed across all facilities to determine the national estimate of pounds of oxytetracycline reduced from discharges as a result of the regulation.

As part of a sampling episode, EPA also performed a preliminary study to

develop a method to measure oxytetracycline in effluent from CAAP facilities. EPA took samples to analyze the effluent from a CAAP facility that produces trout during a time period in which oxytetracycline, in medicated feed, was being used to treat a bacterial infection in some of the animals at the facility. Results of the study indicate that oxytetracycline can be stabilized in samples when preserved with phosphoric acid and maintained below 4 °C prior to analysis. The method found levels of oxytetracycline to range from <0.2 μg/L (which was the method detection limit) in the supply and hatchery effluent to 110 μg/L in the influent to the offline settling basin. The level detected in the combined raceway effluent was 0.95 μg/L. See the analysis report (DCN 63011) for additional information.

H. Environmental Assessment and Benefits Analysis

EPA's environmental assessment and benefits analysis for the proposed rule consisted of two efforts. First, EPA reviewed and summarized literature it had obtained regarding environmental impacts of the aquaculture industry, focusing particularly on segments of the industry in the scope of the proposed rule. Second, EPA used estimates of pollutant loading reductions associated with the proposed requirements to assess improvements to water quality that might arise from the proposed requirements, and monetized benefits from these water quality improvements.

EPA's approach to the environmental assessment and benefits analysis for the final rule is similar to the approach for the proposed rule, except that EPA has incorporated new data, information, and methods that were not available at the time of proposal, particularly those sources described in Section V of this Preamble. For example, literature, discussions, and data submitted by stakeholders both through the public comment process on the proposed rule as well as at other forums were considered. EPA also used facility-specific data provided by or developed from the detailed survey responses. EPA has updated and revised its summary of material relating to environmental impacts of CAAP facilities in Chapter 7 of the Economic and Environmental Benefit Analysis for today's final rule (DCN 63010). EPA's revised benefits analysis are described in both Section X of this Preamble as well as in Chapter 8 of the Economic and Environmental Impact Analysis (DCN 63010).

VII. Who Is Subject to This Rule?

This section discusses the scope of the final rule and explains what wastewaters are subject to the final limitations and standards.

A. Who Is Subject to This Rule?

Today's rule applies to commercial (for-profit) and non-commercial (generally, publicly-owned) facilities that produce, hold or contain 100,000 pounds or more of aquatic animals per year. Any 12 month period would be considered a year for the purposes of establishing coverage under this rule.

While facilities producing fewer than 100,000 pounds of aquatic animals per year are not subject to this rule, in specific circumstances they may require NPDES permits that include limitations developed on a BPJ basis. An aquatic animal production facility producing fewer than 100,000 pounds of aquatic animals per year will be subject to the NPDES permit program if it is a CAAP as defined in 40 CFR 122.24. As explained in the proposed rule, EPA limited the scope of the regulation it was considering to facilities that are CAAPs above this production threshold.

The Agency concluded that facilities below the threshold would likely experience significant adverse economic impacts if required to comply with the proposed limitations. EPA concluded that these smaller CAAP facilities would have compliance costs in excess of 3 percent of revenues. Further, smaller CAAP facilities account for a smaller relative percentage of total CAAP TSS discharges and only limited removals would be obtained from the proposed BPT/BCT/BAT control. 67 FR 57872, 57884. Other types of facilities also not covered by today's action include closed pond systems (most of which do not meet the regulatory definition of a CAAP facility), molluscan shellfish operations, including nurseries, crawfish production, alligator production, and aquaria and net pens rearing native species released after a growing period of no longer than 4 months to supplement commercial and sport fisheries. This last exclusion applies primarily to Alaskan non-profit facilities which raise native salmon for release into the wild in flow-through systems and then hold them for a short time in net pens preceding their release. The flow-through portions of these facilities are within the scope of the rule, if they produce 100,000 pounds or more per year, but the net pen portions would be excluded from regulation. EPA determined for the types of excluded systems or production operations listed above either that they generate minimal pollutant discharges in the baseline or that available pollutant control technologies will reduce pollutant loadings from these operations by only minimal amounts. For further explanation, see the proposal at 67 FR 57572, 57885–86.

Facilities that indirectly discharge their process wastewater (*i.e.*, facilities that discharge to POTWs) are also not subject to today's rule. EPA did not propose and is not establishing pretreatment standards for existing or new indirect sources. As explained above, the bulk of pollutant discharges from CAAP facilities consists of TSS and BOD. POTWs are designed to treat these conventional pollutants. Moreover, CAAP facilities discharge nutrients in concentrations lower in full-flow discharges, and similar in off-line settling basin discharges, to nutrient concentrations found in human wastes discharged to POTWs. EPA has concluded that the POTW removals of TSS would achieve equivalent nutrient removals to those obtained by the options considered for this rulemaking for direct dischargers. EPA, therefore, concluded that there would be no pass through of TSS or nutrients needing regulation. Indirect discharging facilities are still subject to the General Pretreatment Standards (40 CFR 403) and any applicable local limitations. EPA has also determined that there are few indirect dischargers in this industry.

B. What If a Facility Uses More Than One Production System?

EPA has found that several detailed survey respondents are operating more than one type of production system. A facility is subject to the rule if the total production from any of the regulated production systems meets the production threshold. The facility would need to demonstrate compliance with the management practices required for each of the regulated production systems it is operating.

C. What Wastewater Discharges Are Covered?

This rule covers wastewaters generated by the following operations/processes: Effluent from flow-through, recirculating and net pen facilities. The flow-through and recirculating subcategory (Subpart A) applies to wastewaters discharged from these systems.

The type of production system determines the nature, quantity, and quality of effluents from CAAP facilities. Flow-through systems commonly use raceways or tanks and are characterized by continual flows of relatively large volumes of water into and out of the rearing units. Some flow-through systems discharge a single, combined effluent stream with large water volumes and dilute pollutant concentrations. Other flow-through systems have two or more discharge streams, with the process water in which the fish are raised as the primary discharge. This discharge, referred to as raceway effluent or bulk flow, is characterized by a large water volume and dilute pollutant concentrations. The secondary discharges from flow-through systems with multiple discharges result typically from some form of solids settling through an off-line settling basin (OLSB) or other solids removal devices. The discharges from off-line settling basins or solids removal devices have low water volumes and more concentrated pollutants. The supernatant from the OLSB may be discharged through a separate outfall or may be recombined prior to discharge with the raceway effluent.

Recirculating systems may also have two waste streams: Overtopping wastewater and filter backwash. Overtopping is a continuous blowdown from the production system to avoid the buildup of dissolved solids in the production system, and filter backwash is generated by cleaning the filter used to treat the water that is being recirculated back to the production system. Overtopping wastewater is usually small in volume (a fraction of the total system volume on a daily basis) and has higher TSS concentrations than a full flow discharge. Filter backwash wastewater is typically low in volume and is as concentrated as wastewater from similar devices at flow-through systems.

Net pen systems are located in open waters and thus are characterized by the flow and characteristics of the surrounding water body and by the addition of raw materials to the pens including feed, drugs and the excretions from the confined aquatic animals.

VIII. What Are the Requirements of the Final Regulation and the Basis for These Requirements?

This section describes, by subcategory, the options EPA considered and selected as a basis for today's rule. For each subcategory, EPA provides a discussion, as applicable, for the options considered for each of the regulatory levels identified in the CWA (*i.e.*, BPT, BCT, BAT, NSPS). For a detailed discussion of all technology options considered in the development of today's final rule, see the proposal (*see* 67 FR 57872), the NODA (*see* 68 FR 75068) or Chapter 9 of the Technical

Development (TDD) for today's final rule.

Based on the information in the record for the final CAAP rule, EPA has determined that the selected technology for the flow-through and recirculating systems subcategory and the net pens subcategory are technically available. EPA has also determined that the technology it selected as the basis for the final limitations or standards has effluent reductions commensurate with compliance costs and is economically achievable for the applicable subcategory. EPA also considered the age, size, processes, and other engineering factors pertinent to facilities in the scope of the final regulation for the purpose of evaluating the technology options. None of these factors provides a basis for selecting different technologies from those EPA has selected as its technology options for today's rule (*see* Chapter 5 of the TDD for the final rule for further discussion of EPA's analyses of these factors).

As previously explained, EPA adopted a production threshold cutoff as the principal means of reducing economic impacts on small businesses and administrative burden for control authorities associated with the treatment technologies it considered. EPA notes that certain direct dischargers that are not subject to today's effluent limitations or standards will still require a NPDES discharge permit developed on a case-by-case basis if they are CAAPs as defined in 40 CFR 122.24.

The new source performance standards (NSPS) EPA is today establishing represent the greatest degree of effluent reduction achievable through the best available demonstrated control technology. In selecting its technology basis for today's new source performance standards (NSPS), EPA considered all of the factors specified in CWA section 306, including the cost of achieving effluent reductions. EPA used the appropriate technology option for developing today's standards for new direct dischargers. The new source technology basis for both subcategories is equivalent to the technology bases upon which EPA is setting BPT/BCT/BAT (see Chapter 9 of the EEBA). EPA has thoroughly reviewed the costs of such technologies and has concluded that such costs do not present a barrier to entry. The Agency also considered energy requirements and other non-water quality environmental impacts for the new source technology basis and found no basis for any different standards from those selected for NSPS. Therefore, EPA concluded that the NSPS technology basis chosen for both

subcategories constitute the best available demonstrated control technology. For a discussion on the compliance date for new sources, *see* section I.E. of today's final rule.

A. What Technology Options Did EPA Consider for the Final Rule?

Among the options EPA considered for the final rule for flow-through and recirculating systems in addition to the options presented in the proposed rule were (i) establishing no national effluent limitations (ii) establishing limitations and BMPs based on technology options A and B, and (iii) establishing narrative limitations based on BMPs only. Based on analysis presented in the NODA, EPA focused it analysis on these latter three options. For net pens, EPA considered three options: no national requirements, requirements equivalent to those proposed but for new sources only, and essentially the same requirements for existing and new sources as those in the proposed rule.

B. What Are the Requirements for the Flow-Through and Recirculating Systems Subcategory?

The following discussion explains the BPT/BCT/BAT limitations and NSPS EPA is promulgating for flow-through and recirculating system facilities.

1. BPT

After considering the technology options described in the previous section and the factors specified in section 304(b)(1)(B) of the CWA, EPA is establishing nationally applicable effluent limitations guidelines for flow-through and recirculating system CAAP facilities producing 100,000 pounds or more of aquatic animals per year for the reasons noted above at VIII.A.

EPA based the final requirements on production and operational controls that include a rigorously implemented feed management program. Programs of production and operational controls that include feed management systems, proper storage of material and adequate solids controls, and proper operation and maintenance are in wide use at existing flow-through and recirculating system facilities. Based on the detailed survey results, EPA estimates that such programs are currently used at 61 flow-through and recirculating facilities out of 242 total facilities. The costs of effluent removals associated with the evaluated practices are reasonable. The cost per pound of pollutant removed is $2.77 as measured using the higher of the removals for either BOD or TSS at each facility. (The removals for these parameters are not summed because of possible overlap and double counting.)

Based on its review of the data and information it obtained during this rulemaking, EPA has concluded that the key element in achieving effective pollution control at CAAP facilities is a well-operated program to manage feeding, in addition to good solids management. Feed is the primary source of TSS (and associated pollutants) in CAAP systems, and feed management plans are the principal tool for minimizing accumulation of uneaten feed in CAAP wastewater. Excess feed in the production system increases the oxygen demand of the culture water and increases solids loadings. In addition, solids from the excess feed usually settle and are naturally processed with the feces from the fish. Excess feed and feces accumulate in the bottom of flow-through and recirculating systems or below net pens. Ensuring that the aquatic animal species being raised receive the quantity of feed necessary for proper growth without overfeeding, and the resulting accumulation of uneaten feed, is a challenging task. Achieving the optimal feed input requires properly designing a site-specific feeding regimen that considers production goals, species, rearing unit water quality and other relevant factors. It also requires careful observation of actual feeding behavior, good record keeping, and on-going reassessment.

After full examination of the data supporting EPA's model technology, EPA has decided not to establish numerical TSS limitations. While the model technology will effectively remove solids to a very low level, EPA's data show wide variability, both temporally and across facilities, in the actual TSS levels achieved. EPA does not have a record basis for establishing numeric TSS limitations derived from its data set that are appropriate for all sites under all conditions. EPA believes that establishing a uniform numeric TSS limitation would result in requirements that are too stringent at some sites and not stringent enough at others. This is because feed management, while an effective pollution reduction technology for this industry, is not amenable to the same level of engineering process control as traditional treatment technologies used in other effluent guidelines. The basis for this conclusion is further explained below.

Clean Water Act sections 301(b)(1)(A) and 301(b)(2) require point sources to achieve effluent limitations that require the application of the BPT/BCT/BAT selected by the Administrator under section 304(b). Customarily, EPA implements this requirement through the establishment of numeric effluent limitations calculated to reflect the levels of pollutant removals that facilities employing those technologies can consistently achieve. EPA traditionally uses a combination of sampling data and data reported in discharge monitoring reports from well-operated systems employing the model technology to calculate numeric effluent limitations.

In the proposed rule and the NODA, EPA used a similar approach to calculate numeric effluent limitations for TSS from a partial data set composed of well operated CAAP facilities employing a combination of wastewater treatment and management practices to reduce TSS concentrations in the discharged effluent. To reduce TSS discharge levels, the facilities examined by EPA used settling ponds and a number of different techniques, including feed management programs and periodic solids removal from both the culture water and settling ponds.

EPA's examination of well-operated facilities also identified several facilities using feed management and other operational and management controls alone that were achieving the same low levels of TSS discharge as facilities using settling ponds in combination with good feed management.

Based on EPA's examination of the data in its record, the Agency has concluded that a combination of settling technology and feed management control practices or rigorous feed management control and proper solids handling practices alone will achieve low levels of TSS. Operational measures like a feed management system, however, are not technologies that reflect the same degree of predictability as can be expected from wastewater treatment technology based on chemical or other physical treatment. While EPA is confident that its chosen technology can consistently achieve BPT treatment levels of solids removal, the Agency recognizes that feed management systems may not have the precision or consistently predictable performance from site to site that come with the traditional wastewater treatment technologies. The record confirms that there is variability in results associated with the use of feed management systems and other operational measures to control solids. Thus, EPA determined that it should not establish specific numeric TSS limitations based on the model technology. This conclusion is supported by a number of commenters who maintained that consistently achieving the proposed TSS levels would require installation of additional settling treatment structures, with little additional environmental benefit.

EPA's decision not to set uniform numeric TSS limitations based on rigorous feed management and good solids management is further supported by its analysis of measured or predicted TSS concentrations at facilities employing this technology. EPA's effluent monitoring data show differences in the measured TSS concentration in discharges at facilities employing feed management programs from the predicted TSS concentration levels derived using EPA's calculation from the data on feed used at BPT/BAT facilities. For this comparison, EPA calculated a TSS concentration that could be achieved through feed management plans using the data on feed and fish production at surveyed facilities. EPA then compared these concentrations, where available, with the actual TSS levels reported by those facilities in their discharge monitoring reports. The differences between the calculated TSS levels and reported levels may result from differences in application of feed management practices, variation in the flows or dilution of the effluent.

EPA recognizes that it would be feasible to calculate numeric effluent limitations for TSS based on treatment technologies alone, *i.e.*, eliminating best management practices from the technology basis for today's rule. EPA did not employ this approach for three reasons. First, EPA has determined that primary treatment in the form of quiescent zones in the culture water tanks and settling ponds by themselves are not the best technology available for treating TSS. Instead, rigorous feed management in conjunction with good solids handling practices constitutes a better technology for controlling this pollutant. Second, EPA is concerned that establishing numeric limitations for TSS based on primary and secondary settling may not be a practicable technology. Commenters pointed out that site and land availability constraints might limit their ability to install the additional treatment needed to achieve TSS limitations. Third, EPA believes based on its analysis of the data, that comparable discharge levels can be achieved using feed management and other management practices alone as can be achieved using these practices in combination with settling technologies. Thus, while settling technology may be amenable to more precise control, EPA believes that the overall environmental benefits of this technology relative to rigorous feed and solids handling management alone are negligible.

EPA is further concerned that establishing a numeric limit for TSS

could provide an incentive for facilities to achieve the limit through dilution and would not reduce the pollutant loads discharged to receiving streams. While dilution is generally prohibited as a means of achieving effluent limitations, this prohibition is harder to enforce at CAAP facilities than in most other systems because the flow of culture water is dependent on a wide range of factors and is highly variable from one facility to another. Thus it would be impossible for regulatory authorities to determine if water use was being manipulated to dilute TSS concentration. Due to variations in water use from facility to facility, EPA also decided not to establish mass-based numeric TSS limitations on a national basis. Solids control operational measures such as feed management and the requirement to focus on the proper operation of existing solids control structures are expected to achieve reductions in the TSS concentrations and at the same time reduce the TSS loadings being discharged. This approach is supported by DMR data from facilities in Idaho which have had to comply with feed management BMP requirements in their general permit. This data demonstrates that improved performance can be achieved through BMPs (DCN 63012). A comparison of DMR data from Idaho prior to the issuance of a general permit in calendar year 1999 with data following compliance with the general permit indicates that 64 percent of the facilities have reduced the TSS loads discharged from the facility with an average TSS reduction of 75 percent.

For these reasons, EPA has expressed effluent limitations in this rule in the form of narrative standards, rather than as numeric values. EPA has a legal authority to do so. The CWA defines "effluent limitation" broadly, and EPA's regulations reflect this as well. Each provides that an effluent limitation is "*any restriction*" imposed by the permitting authority on quantities, discharge rates and concentrations of a pollutant discharged into a water of the United States. CWA section 502(11) (emphasis supplied); 40 CFR 122.2 (emphasis supplied). Neither definition requires an effluent limitation to be expressed as a numeric limit. The DC Circuit observed, "Section 502(11) defines 'effluent limitation' as '*any* restriction' on the amounts of pollutants, not just a numerical restriction." *NRDC* v. *EPA*, 673 F.2d 400, 403 (DC Cir.) (emphasis in original), *cert. denied sub nom. Chemical Mfrs. Ass'n* v. *EPA*, 459 U.S. 879 (1982). In short, the definition of "effluent limitation" is not limited to a single type of restriction, but rather contemplates a range of restrictions that may be used as appropriate. EPA has concluded that it is appropriate to express today's BPT/BCT/BAT limitations in non-numeric form. These narrative limitations reflect a technology demonstrated to achieve effective solids removals while still giving facilities flexibility in determining how to meet them.

Today's BPT regulation requires CAAP facilities to comply with specified operational and management requirements—best management practices (BMPs)—that will minimize the generation and discharge of solids from the facility. These requirements are non-numeric effluent limitations based on the technologies EPA has determined are BPT.

The final regulation requires adoption of specified solids control practices. *See, e.g.*, § 451.11(a) and § 451.21(a). Thus, to control the discharge of solids from flow-through and recirculating system facilities, the final rule requires minimizing the discharge of uneaten feed through a feed management program. *See* § 451.11(a) of this rule. Complying with this limitation will require a CAAP facility to identify feeding practices which optimize the addition of feed to achieve production goals while minimizing the amount of uneaten feed leaving the rearing unit. Such a program should include practices such as periodic calibration of automatic feeders, visual observation of feeding activity and discontinuation of feeding when the animals stop eating. The rule also requires that CAAPs maintain records of feed inputs and estimates of the numbers and weight of aquatic animals in order to calculate representative feed conversion ratios. *See* § 451.11(a)(1) of this rule. Development of feed conversion ratios is a key component in a properly functioning feed management system because it allows the facility to calibrate more accurately the feeding needs of the species being raised. This, in turn, will result in further improvement in control of solids at the operation.

In addition to feed management, EPA also requires flow-through and recirculating system facilities to identify and implement procedures for routine cleaning. *See* § 451.11(a)(2). This will ensure that CAAP facilities develop practices to minimize the build-up and subsequent discharge of solids from the rearing units. The facility must also identify procedures with respect to harvesting, inventorying and grading of fish so as to minimize disturbance and

discharge of solids from the facility during these activities.

The final rule also provides that facilities must remove dead fish and fish carcasses from the production system on a regular basis and dispose of them to avoid the discharge to waters of the U.S. § 451.11(a)(3). EPA is establishing an exception to this requirement when the permit writer authorizes a discharge to benefit the aquatic environment. The following example explains one circumstance in which a permit writer could authorize such a discharge. There are a number of federal, state, and tribal hatcheries that are raising fish for stocking or mitigation purposes. In some cases, these facilities have been approved to discharge fish carcasses along with the live fish that are being stocked. In these situations, the carcasses are serving as a source of nutrients and food to the fish being stocked in these waters. The exception would apply in these circumstances if the permitting authority determines that the addition of fish carcasses to surface water will improve water quality.

Facilities must also implement measures that address material storage and structural maintenance. In the case of material storage, EPA is requiring facilities to identify and develop practices to prevent inadvertent spillage of drugs, pesticides, and feed from the facility. § 451.11 (b). This would include proper storage of these materials. EPA is also requiring facilities to identify proper procedures for cleaning, containing and disposing of any spilled material. EPA's assessment, based on site visits and sampling visits, indicates that facilities may have varying degrees of spill prevention procedures and containment and structural maintenance practices to address these requirements.

The final rule also includes a requirement that facilities inspect and provide regular maintenance of the production system and the wastewater treatment system to ensure that they are properly functioning. § 451.11(c). One area of concern addressed by this requirement is the potential accumulation of solids (especially large solids such as carcasses and leaves) that could clog screens that separate the raceway from the quiescent zone. These solids could prevent the flow of water through the screen causing water to instead flow over the screen and impair the passage of solids into the quiescent zone. Proper maintenance should ensure that screens are regularly inspected and cleaned.

The final rule also requires that facilities conduct routine inspections to identify any damage to the production system or wastewater treatment system

and that facilities repair this damage promptly. EPA has not specified any design requirement for structural components of the CAAP facility. Rather, it has adopted the requirement that facilities identify practices that will ensure existing structures are maintained in good working order. Flow-through and recirculating facilities are also required to keep records as described previously and to conduct routine training for facility staff on spill prevention and response.

As discussed further below, in the final rule, EPA is not establishing numeric limits for any drug or pesticide but is requiring CAAP facilities to ensure proper storage of drugs, pesticides and feed to prevent spills and any resulting discharge of spilled drugs and pesticides. EPA is also establishing a requirement to implement procedures for responding to spills of these materials to minimize their discharge from the facility. *See* § 451.11(c)(2) of this rule. Facilities must also train their staff in spill prevention and proper operation and cleaning of production systems and equipment. *See* § 451.11(e) of this rule. The detailed survey did not provide information about spill prevention, but during site visits and sampling visits EPA identified containment systems and practices. EPA's site visit information indicated that CAAP facilities currently employ a number of different measures to prevent spills and some have established in-place systems to address spills in the event they occur. The effect of this narrative limitation will be to promote increased care in the handling of these materials. Its adoption as a regulatory requirement provides an additional incentive for facility operators currently employing effective spill control measures to continue such practices when handling drugs and pesticides. Moreover, because EPA has adopted the same requirements for existing and new sources (*see* discussion below), this will ensure that new sources employ the same highly protective measures as existing sources have employed successfully to protect against spills.

Today's regulation does not include any requirements specifically addressing the release of non-native species. The final regulation, however, includes a narrative effluent limitation that requires facilities to implement operational controls that will ensure the production facilities and wastewater treatment structures are being properly maintained. Facilities must conduct routine inspections and promptly repair damage to the production systems or wastewater treatment units. This requirement, described in more detail in

Section VI.D., will aid in preventing the release of various materials, including live fish.

2. BAT

EPA is establishing BAT at a level equal to BPT for the flow-through and recirculating system discharge subcategory. For this subcategory, EPA did not identify any available technologies that are economically achievable for the subcategory that would achieve more stringent effluent limitations than those considered for BPT. Because of the nature of the wastes generated from CAAP facilities, advanced treatment technologies or practices to remove additional toxic or nonconventional pollutants that would be economically achievable on a national basis do not exist beyond those already considered.

3. BCT

EPA evaluated conventional pollutant control technologies and did not identify a more stringent technology for the control of conventional pollutants for BCT limitations that would be affordable than the final requirements considered. Other technologies for the control of conventional pollutants include biological treatment, but this technology is not affordable for the subcategory as a whole. Consequently, EPA has not promulgated BCT limitations or standards based on a different technology from that used as the basis for BPT limitations and standards.

4. NSPS

After considering the technology options described in the proposal and NODA and evaluating the factors specified in section 306 of the CWA, EPA is promulgating standards of performance for new sources equal to BPT, BAT, and BCT. There are no more stringent technologies available for NSPS that would not represent a barrier to entry for new facilities, *see* Section IX for more discussion of the barrier to entry analysis. Because of the nature of the wastes generated in CAAP facilities, EPA has not identified advanced treatment technologies or practices to remove additional solids (*e.g.*, smaller particle sizes) in TSS or other pollutants that would be generally affordable beyond those already considered.

EPA determined that NSPS equal to BAT will not present a barrier to entry. The overall impacts from the effluent limitations guidelines on new sources would not be any more severe than those on existing sources. This is because the costs faced by new sources are generally the same as, or lower than,

those faced by existing sources. It is generally less expensive to incorporate pollution control equipment into the design at a new facility than it would be to retrofit the same pollution control equipment in an existing plant. At a new facility, no demolition is required and space constraints (which can add to retrofitting costs if specifically designed equipment must be ordered) may be less of an issue.

C. What Are the Requirement for the Net Pen Subcategory?

The following discussion explains the BPT/BAT/BCT limitations and NSPS EPA is promulgating for Net Pen Systems.

1. BPT

After considering the technology options described in the proposal and the factors specified in Section 304(b)(1)(B) of the Clean Water Act, EPA is establishing nationally applicable effluent limitations for net pen facilities producing 100,000 pounds or more of aquatic animals per year. Today's BPT regulations requires CAAP net pen systems, like CAAP flow-through and recirculating systems, to comply with specified operational practices and management requirements. These requirements are non-numeric effluent limitations based on technologies EPA has evaluated and determined are cost-reasonable, available technologies.

Based on the detailed survey results, EPA estimates that such programs are currently in use at most or all the net pen systems. As a result, the cost to facilities of meeting the BPT requirements is very low. To EPA's knowledge, all existing net pen facilities that are currently covered by NPDES permits are subject to permit requirements comparable to today's limitations. Therefore, EPA concludes that the BPT limits are both technically available and cost reasonable for the net pen subcategory.

EPA rejected the establishment of numeric effluent limitations for net pens for obvious reasons. Because of the nature of the facilities, net pens cannot use physical wastewater control systems except at great cost. Located in open waters, nets are suspended from a floating structure to contain the crop of aquatic animals. Nets are periodically changed to increase the mesh size as the fish grow in order to provide more water circulating inside the pen. The pens are anchored to the water body floor and sited to benefit from tidal and current action to move wastes away from, and bring oxygenated water to, the pen. As a result, these CAAP facilities experience a constant in- and out-flow

of water. Development of a system to capture the water and treat the water within the pen would be prohibitively expensive. EPA, therefore, rejected physical treatment systems as the basis for BPT limitations. Instead, EPA is promulgating narrative effluent limitations.

As was the case with flow-through and recirculating systems, feed management programs are a key element of the promulgated requirements for the reasons explained above and in the proposal at 67 FR 57872, 57887. Consequently, for the control of solids, the final regulation requires that net pen CAAP facilities minimize the accumulation of uneaten feed beneath the pen through the use of active feed monitoring and management practices. § 451.21(a). These strategies may include either real-time monitoring (*e.g.*, the use of video monitoring, digital scanning sonar, or upweller systems); monitoring of sediment quality beneath the pens; monitoring of the benthic community beneath the pens; capture of waste feed and feces; or the adoption of other good husbandry practices, subject to the permitting authority's approval.

As noted, feed management systems are effective in reducing the quantity of uneaten feed. Facilities should limit the feed added to the pens to the amount reasonably necessary to sustain an optimal rate of fish growth. In determining what quantity of feed will result in minimizing the discharge of uneaten feed while at the same time sustaining optimal growth, a facility should consider, among others, the following factors: The types of aquatic animals raised, the method used to feed the aquatic animals, the facility's production and aquatic animal size goals, the species, tides and currents, the sensitivity of the benthic community in the vicinity of the pens, and other relevant factors. In some areas, deep water and/or strong tides or currents may prevent significant accumulation of uneaten feed such that active feed monitoring is not needed. Several states with significant numbers of net pens (*e.g.*, Washington, Maine) already require feed management practices, which may include active feed monitoring, to minimize accumulation of feed beneath the pens. Facilities will need to ensure that whatever practices they adopt are consistent with the requirements of their state NPDES program.

In order to implement a feed management system, the facility must also track feed inputs by maintaining records documenting feed and estimates of the numbers and weight of aquatic animals in order to calculate representative feed conversion ratios. § 451.21(g). As previously explained, development of feed conversion ratios are a necessary element in any effective feed management system.

Real-time monitoring represents a widely-used business practice that is employed by many salmonid net pen facilities to reduce feed costs. Net pen systems do not present the same opportunities for solids control as do flow-through or recirculating systems for the obvious reason that ocean water is continuously flowing in and out of the net pens. Therefore, in EPA's view, feed monitoring, including real time monitoring and other practices is an important and cost reasonable practice to control solids discharges.

The final rule includes a narrative limitation requiring CAAP net pen facilities to collect, return to shore, and properly dispose of all feed bags, packaging materials, waste rope and netting. § 451.21(b). This will require that net pen facilities have the equipment (e.g., trash receptacles) to store empty feed bags, packaging materials, waste rope and netting until they can be transported for disposal. EPA is also requiring that net pens minimize any discharges associated with the transporting or harvesting of fish, including the discharge of blood, viscera, fish carcasses or transport water containing blood. § 451.21(c). During stocking or harvesting of fish, some may die. The final limitations require facilities to remove and dispose of dead fish properly on a regular basis to prevent discharge. Discharge of dead fish represents an environmental concern because they may spread disease and attract predators, which could imperil the structural integrity of the containment system. The wastes and wastewater associated with the transport or harvest of fish have high BOD and nutrient concentrations and should be disposed of at a location where they may be properly treated.

The final regulations also require net pen facilities to ensure the proper storage of drugs, pesticides, and feed to avoid spilling these materials and subsequent discharge. See § 451.21(e)(1) of this rule. Facilities must also implement procedures for properly containing, cleaning and disposing of any spilled material. See § 451.21(e)(2) of this rule. As previously discussed, excess feed may present a number of different environmental problems. Preventing spills of feed is consequently important. Additionally, net pens may use different pesticides and drugs in fish production. Preventing their release is similarly important. The final regulation also includes a narrative

limitation, similar to that for CAAP flow-through and recirculating systems, requiring that net pen facilities adequately train facility personnel in how to respond to spills and proper clean-up and disposal of spilled material. See § 451.21(h) of this rule.

Next, the final regulation requires regular inspection and maintenance of the net pen § 451.21(f). This would include any system to prevent predators from entering the pen. Net pens are vulnerable to damage from predator attack or accidents that result in the release of the contents of the nets, including fish and fish carcasses. Given the economic incentive to prevent the loss of production, EPA assumes facilities will conduct routine inspections of the nets to ensure they are not damaged and make repairs as soon as any damage is identified. Most net pen facilities are already doing these inspections. However, in evaluating this technology option, EPA estimated costs for increased inspections at every net pen facility in order to ensure that costs are not underestimated.

Like the final BPT limitations for flow-through and recirculating systems, the BPT limitations for net pens do not include any requirements specifically addressing the release of non-native species. The final regulation, however, includes a narrative effluent limitation that requires facilities to implement operational controls that will ensure the production facilities and wastewater treatment structures are being properly maintained. Facilities must conduct routine inspections and promptly repair damage to the production systems or wastewater treatment units. EPA included this requirement to ensure achievement of the other BPT limitations for net pens such as the prohibition on the discharge of feed bags, packaging materials, waste rope and netting at net pens, and the requirement to minimize release of solids, fish carcasses and viscera. This requirement will also aid in preventing the release of other materials including live fish.

2. BAT

EPA is establishing BAT at a level equal to BPT for the net pen subcategory. For this subcategory, EPA did not identify any available technologies that are economically achievable that would achieve more stringent effluent limitations than those considered for BPT. Because of the nature of the wastes generated from CAAP net pen facilities, EPA did not identify any advanced treatment technologies or practices to remove additional toxic and nonconventional

pollutants that would be economically achievable on a national basis beyond those already considered.

3. BCT

EPA evaluated conventional pollutant control technologies and did not identify a more stringent technology for the control of conventional pollutants for BCT limitations than the final requirements considered. Consequently, EPA has not promulgated BCT limitations or standards based on a different technology from that used as the basis for BPT limitations and standards.

4. NSPS

After considering the technology requirements described previously under BPT, and the factors specified in section 306 of the CWA, EPA is promulgating standards of performance for new sources equal to BPT, BAT, and BCT. There are no more stringent best demonstrated technologies available. Because of the nature of the wastes generated and the production system used, EPA has not identified advanced treatment technologies or practices that would be generally affordable beyond those already considered.

Although siting is not specifically addressed with today's standards, proper siting of new facilities is one component of feed management strategies designed to minimize the accumulation of uneaten feed beneath the pens and any associated adverse environmental effects. When establishing new net pen CAAP facilities, consideration of location is critical in predicting the potential impact the net pen will have on the environment. Net pens are usually situated in areas which have good water exchange through tidal fluctuations or currents. Good water exchange ensures good water quality for the animals in the nets. It also minimizes the concentration of pollutants below the nets. In implementing today's rule for new net pen operations, facilities and permit authorities should give careful consideration to siting prior to establishing a new net pen facility.

EPA has concluded that NSPS equal to BAT does not present a barrier to entry. The overall impacts from the effluent limitations guidelines on new source net pens are no more severe than those on existing net pens. The costs faced by new sources generally should be the same as, or lower than, those faced by existing sources. It is generally less expensive to incorporate pollution control equipment into the design at a new facility than it is to retrofit the same pollution control equipment in an existing facility.

Although EPA is not establishing standards of performance for new sources for small cold water facilities (*i.e.*, those producing between 20,000 and 100,000 pounds of aquatic animals per year), such facilities would be subject to existing NPDES regulations and BPT/BAT/BCT permit limits developed using the permit writer's "best professional judgment" (BPJ). EPA, based on its analysis of existing data, determined that new facilities would most often produce 100,000 pounds of aquatic animals or more per year because of the expense of producing the aquatic animals. Generally, the species produced are considered of high value and are produced in such quantities to economically justify the production. For example, one net pen typically holds 100,000 pounds of aquatic animals or more. In reviewing USDA's Census of Aquaculture and EPA's detailed surveys, EPA has not identified any existing commercial net pen facilities producing fewer than 100,000 pounds of aquatic animals per year.

Offshore aquatic animal production is an area of potential future growth. As these types of facilities start to produce aquatic animals, those with 100,000 pounds or more per year will be subject to the new source requirements established for net pens as well as NPDES permitting.

D. What Monitoring Does the Final Rule Require?

The final rule does not require any effluent monitoring. In the case of net pen facilities, however, it does require CAAPs to adopt active feed monitoring and management practices that will most often include measures to observe the addition of feed to the pen. Net pen facilities subject to today's rule must develop and implement active feed monitoring and management strategies to minimize the discharge of solids and the accumulation of uneaten feed beneath the pen. Many existing net pen facilities use a real-time monitoring system such as video cameras, digital scanning sonar, or upweller systems to accomplish this. With a real-time monitoring system, when uneaten feed is observed falling beneath the pen feeding should stop. Depending on the location and other site-specific factors at the facility, a facility may adopt other measures in lieu of real time monitoring. These may include monitoring of sediment or the benthic community quality beneath the pens, capture of waste feed and feces or other

good husbandry practices that are approved by the permitting authority.

E. What Are the Final Rule's Notification, Recordkeeping, and Reporting Requirements?

The final rule establishes requirements for reporting the use of spilled drugs, pesticides or feed that result in a discharge to waters of the U.S. by CAAP facilities. This provision ensures that, any release of spilled drugs, pesticides and feed to waters of the U.S. are reported to the permitting authorities to provide them with necessary information for any responsive action that may be warranted. This will allow regulatory authorities to reduce or avoid adverse impacts to receiving waters associated with these spills. EPA is requiring that any spill of material that results in a discharge to waters of the U.S. be reported orally to the permitting authority within 24 hours of its occurrence. A written report shall be submitted within 7 days. Facilities are required to report the identity of the material spilled and an estimated amount.

EPA is retaining for the final rule the proposed requirement that CAAP facilities report to the Permitting Authority whenever they apply certain types of drugs under the following conditions. First, the permittee must report drugs prescribed by a veterinarian to treat a species or a disease when prescribed for a use which is not an FDA-approved use (referred to as "extralabel drug use") as described further below. Second, the permittee must report drugs being used in an experimental mode under controlled conditions, known as Investigative New Animal Drugs (INADs). In EPA's view, notifying the Permitting Authority is necessary to ensure that any potential risk to the environment resulting from the use of these drugs can be addressed with site-specific remedies where appropriate. EPA strongly encourages reporting prior to use where feasible, as this provides the Permitting Authority with the opportunity to monitor or control the discharge of the drugs while the drugs are being applied. EPA has not made this an absolute requirement, however, in recognition of the fact that swift action on the part of veterinarians and operators is sometimes necessary to respond to and contain disease outbreaks.

The reporting requirement applies to the permittee and imposes no obligation on the prescribing veterinarian. The reporting requirement for extralabel drug use is not in any way intended to interfere with veterinarians' authority to

prescribe extralabel drugs to treat aquatic animals or other animals in accordance with FFCDA and 40 CFR Part 530. This reporting requirement is promulgated to ensure that permitting authorities are aware of the use at CAAPs of extralabel drugs when such use may result in the release of the drug to waters of the U.S. Because the use is likely to involve adding the drug directly to the rearing unit, EPA believes there is a probability that these drugs may be released to waters of the U.S..

The regulation requires that a permittee must provide a written report to the permitting authority within seven days of agreeing to participate in an INAD study and an oral report preferably in advance of use, but in no event later than seven days after starting to use the INAD. The first written report must identify the drug, method of application, the dosage and what it is intended to treat. The oral report must also identify the drug, method of application, and the reason for its use. Within 30 days after the use of the drug at the facility, the permittee must provide another written report to the permitting authority describing the drug, reason for treatment, date and time of addition, method of addition and total amount added.

EPA has similar reporting requirements for extralabel drug use except that EPA is not requiring a written report in advance of use.

The reporting requirement applies only to those drugs that have not been previously approved for their intended use. Reporting would not be required for EPA registered pesticides and FDA approved drugs for aquatic animal uses when used according to label instructions. Reporting would only be required for INAD drugs and drugs prescribed by a veterinarian for extralabel uses. Because these classes of drugs have not been fully evaluated by FDA for the potential environmental consequences of the use being made of them EPA considers reporting ensures the permitting authority has enough information to make an informed response if environmental problems do occur. EPA has included an exception to the reporting requirement for cases where the INAD or extralabel drug has already been approved under similar conditions for use in another species or to treat another disease and is applied at a dosage that does not exceed the approved dosage. The requirement that the use be under similar conditions is intended to limit the exception to cases where the INAD or extralabel drug use would be expected to produce significantly different environmental impacts from the previously approved use. For example, use of a drug that had been previously approved for a freshwater application, as an INAD in a marine setting would not be considered a similar condition of use, since marine ecosystems may have markedly different vulnerabilities than freshwater ecosystems. Similarly, the use of a drug approved to treat terrestrial animals used as an INAD or extralabel drug to treat aquatic animals would not be considered a similar condition of use. In contrast, the use of a drug to treat fish in a freshwater system that was previously approved for a different freshwater species would be considered use under similar conditions. EPA has concluded that when a drug is used under similar conditions it is unlikely that the environmental impacts would be different than those that were already considered in the prior approval of the drug.

The reporting requirements with respect to INADs are not burdensome. FDA regulations require that the sponsor of a clinical investigation of a new animal drug submit to the Food and Drug Administration certain information concerning the intended use prior to its use. Therefore, this information will be readily available to any CAAP facility that participates in an INAD investigation. Having advance information will enable the permitting authority to determine whether restrictions should be imposed on the release of such drugs.

EPA is also requiring all CAAP facilities subject to today's regulation to develop and maintain a Best Management Practices plan on site. This plan must describe how the permittee will achieve the required narrative limitations. The plan must be available to the permitting authority upon request. Upon completion of the plan, the permittee must certify to the permitting authority that a plan has been developed.

The proposal included a requirement to implement escape prevention practices at facilities where non-native species are being produced. EPA received comments supporting such controls to prevent the release of non-native species. EPA also received comments arguing against controls in this regulation because other authorities are already dealing with non-native species, and because of the complexities of determining what is a non-native species and when such species may become invasive. For example, species raised by Federal and State authorities for stocking may not be "native," but would not generally impose a threat if escapes occurred.

Today's regulation does not include any requirements specifically addressing the release of non-native species. The regulation, however, includes a requirement for facilities to develop and implement BMPs to ensure the production and wastewater treatment systems are regularly inspected and maintained. Facilities are required to conduct routine inspections and perform repairs to ensure proper functioning of the structures. EPA included this requirement to promote achievement of BPT/BAT limitations on the discharge of feed bags, packaging materials, waste rope and netting at net pens, and on the discharge of solids, including fish carcasses and viscera at all facilities. This requirement, described in more detail in Section VI.D, will also aid in preventing the release of other materials, including live fish.

The final regulation also includes a requirement for facilities to report failures and damage to the structure of the aquatic animal containment system leading to a material discharge of pollutants. EPA realizes that most CAAP facilities take extensive measures to ensure structural integrity is maintained. Nonetheless, failures do occur with potentially serious consequences to the environment. The failure of the containment system can result in the release of sediment, fish and fish carcasses which, depending on the magnitude of the release, can have significant impacts on the environment. For net pen systems, failures include physical damage to the predator control nets or the nets containing the aquatic animals, which result in a discharge of the contents of the nets. Damage includes abrasion, cutting or tearing of the nets and breakdown of the netting due to rot or ultra-violet exposure. For flow-through and recirculating systems, a failure includes a collapse or damage of a rearing unit or wastewater treatment structure; damage to pipes, valves, and other plumbing fixtures; and damage or malfunction to screens or physical barriers in the system, which would prevent the unit from containing water, sediment, and the aquatic animals. In the event of a reportable failure as defined in the NPDES permit, EPA is requiring CAAP facilities to report to the permit authority orally within 24 hours of discovering a failure and to follow the oral report with a written report no later than seven days after the discovery of the failure. The oral report must include the cause of the failure and the materials that have likely been released. The written report must include a description of the cause of the failure,

the time elapsed until the failure was repaired, an estimate of the types and amounts of materials released and the steps that will be taken to prevent a recurrence. Because the determination of what constitutes damage resulting in a "material" discharge varies from one facility to the next, EPA encourages permitting authorities to include more specific reporting requirements defining these terms in the permit. Such conditions might recognize variations in production system type and environmental vulnerability of the receiving waters.

Today's regulation requires record-keeping in conjunction with implementation of a feed management system. As previously explained, EPA is requiring flow-through, recirculating and net pen CAAP facilities subject to today's regulation to keep records on feed amounts and estimates of the numbers and weight of aquatic animals in order to calculate representative feed conversion ratios. The feed amounts should be measured at a frequency that enables the facility to estimate daily feed rates. The number and weight of animals contained in the rearing unit may be recorded less frequently as appropriate.

Flow-through and recirculating facilities subject to today's requirements must record the dates and brief descriptions of rearing unit cleaning, inspections, maintenance and repair. Net pen facilities must keep the same types of feeding records as described above and record the dates and brief descriptions of net changes, inspections, maintenance and repairs to the net pens.

IX. What Are the Costs and Economic Impacts Associated With This Rule?

This section discusses the costs and economic impact of the rule promulgated today.

A. Compliance Costs

The information below describes the rule's costs and how EPA determined these costs. A more detailed discussion of how EPA estimated compliance costs is included in the Technical Development Document (EPA–821–R–04–012) and the discussion of the economic impacts is included in the Economic and Environmental Benefits Analysis report (EPA–821–R–04–013). Both of these documents can be found on EPA's Web site, *www.epa.gov/ost/guide/aquaculture.*

1. How Did EPA Estimate the Costs of Compliance With the Final Rule?

EPA estimated costs associated with regulatory compliance for the options it considered to determine the economic impact of the effluent limitations guidelines and standards on the aquaculture industry. The economic impact is a function of the estimated costs of compliance to achieve the requirements. These costs may include initial fixed and capital costs, as well as annual operating and maintenance (O&M) costs. Estimation of these costs began by identifying the practices and technologies that could be used as a basis to meet particular requirements. EPA estimated compliance costs for each facility, based on the specific configuration of the facility as provided in the detailed survey and the implementation of the practices or technologies to meet particular requirements.

EPA developed cost estimates for capital, land, annual O&M, and one-time fixed costs for the implementation of the different best management practices and treatment technologies targeted under the regulatory options. EPA developed the cost estimates from information collected from the detailed survey, site visits, sampling events, published information, vendor contacts, industry comments, and engineering judgment. EPA estimates compliance costs in 2001 dollars that it converted to 2003 dollars using the Engineering News Record construction cost index. All costs presented in this section are reported in pre-tax 2003 dollars, unless otherwise indicated.

The final regulation requires facilities to adopt various management practices to control pollutant discharges and incorporate these practices in a BMP plan. The detailed survey provided information on the use of BMPs at each surveyed facility. In its analyses, EPA estimated the costs associated with implementing various types of BMPs. As explained above, EPA has concluded that BMPs are an effective tool for controlling pollutant discharges. EPA assumed no additional costs for compliance for a facility for particular BMPs when the facility indicated that it had comparable BMPs in place, or EPA found strong evidence that such BMPs were already being implemented at the facility. For example, facilities reporting the use of drugs and pesticides that are located in Washington or Idaho were not costed for drug and pesticide BMPs because the general permits in these states require facilities to implement BMPs related to drugs and pesticides that are at least as stringent as these required by today's rule.

EPA is requiring each facility to develop a BMP plan that describes the practices and strategies it is using to comply with narrative limitations addressing solids control, including

feed management, materials storage (*i.e.*, spill containment), structural maintenance, recordkeeping, and training. For net pen facilities, the BMP plan must also document provisions for complying with narrative limitations related to waste collection and disposal, minimization of discharges associated with transport or harvest, and carcass removal. EPA found that the net pen facilities responding to the detailed survey generally have operational measures in place that address these requirements.

The costs associated with BMP plan development include a one-time labor cost of 40 hours for management staff training and time to develop and write the plan. The plan that EPA costed included time for the manager to (1) identify all waste streams, wastewater structures, and wastewater and manure treatment structures at the site, (2) identify and document standard operating procedures for all BMPs used at the facility, and (3) define management and staff responsibilities for implementing the plan. EPA assumed that each employee at a facility would incur a one time cost of 4 hours for initial BMP plan review. EPA included an annual cost for four hours of management labor to maintain the plan and eight hours of management labor and 4 hours for each employee for training and an annual review of BMP performance. EPA included the cost of developing solids control, spill prevention, and structural maintenance components of the BMP plan in the estimates for all appropriate facilities. EPA also included recordkeeping and training costs as a part of annual operation and maintenance activities for the BMP components.

One part of the solids control component of the BMP plan is feed management. Based on feed and production data reported in the surveys, EPA evaluated the effectiveness of a facility's feed management programs. EPA calculated feed conversion ratios (FCRs) using pounds of feed per pound of live product. These calculated FCRs were compared for groups of facilities (*i.e.*, combinations of ownership, species and production system types such as commercial trout flow-through facilities or government salmon flow-through facilities). EPA found a wide range of FCRs (reported by facilities in their detailed surveys, which were validated by call backs to the facility) among apparently similar facilities within ownership-species-production system groupings.

For example, EPA had good data for 24 of 60 government trout producers using flow-through systems. They

reported a range of FCRs of 0.79 to 1.80 with a median FCR of 1.30. If an individual facility's reported FCR was significantly greater than the median, EPA further evaluated the facility to ascertain the reason for the higher FCR. Facilities that produce larger fish, such as broodstock, might have higher FCRs because the larger fish produce less flesh per unit of food. Facilities with fluctuating water temperatures could also be less efficient than facilities with constant water temperatures. EPA assumed facilities lacking evidence of good feed management practices (based on the calculated FCR) would incur additional costs to improve or establish them. However, EPA did not apply costs for feed management BMPs for facilities with reasonable explanations for the higher FCRs because EPA assumed such facilities were already optimizing feed input or would be able to do so at reasonable cost.

EPA evaluated facilities that did not report FCRs or provide enough data for an estimate by assigning each facility a random FCR between the first and third quartiles of the FCR distribution of the group of facilities (*i.e.*, combinations of ownership, species, and production systems) where it was classified. For its analysis, EPA estimated target FCRs for each group as the 25th percentile value of the category. EPA used these target FCRs in its costing and loadings analyses, but does not intend to set any specific FCR targets at facilities (*see* DCN 62467). These facilities were assigned costs associated with feed management BMPs in the same manner as facilities with calculated FCRs.

Costs for the feed management BMP component include staff time for recordkeeping for feed delivery and daily feeding observations. Management activities associated with the feed management practices were weekly data reviews of feeding records, regular estimates of changes to feeding regimes for each group of aquatic animals, and staff consultations about feeding. For facilities that reported using drugs or pesticides, EPA evaluated costs for (1) storage containment, (2) spill prevention planning and training, and (3) reporting of INAD and extralabel drug uses. For storage containment, EPA evaluated the amount of product stored onsite and estimated containment structure costs specifically for the facility. This capital cost was for the purchase of commercially available drum storage units and pesticide cabinets that will contain spills in the event of leakage or accidental spills. EPA also estimated the costs for management to develop a spill prevention plan, which is included in the facility BMP plan, and annual staff training at the facility (8 hours/year for managers and 4 hours/year for each employee). EPA assumed that reporting to the appropriate regulatory authority would occur 6 times per year for facilities reporting using INAD or extralabel drug uses. The reporting for each occurrence includes 20 minutes for an oral report and 1 hour for a written report. EPA considers these costing assumptions to be conservative and may overstate actual reporting frequency.

In addition, EPA estimated costs for inspections in order to maintain the structural integrity of the aquatic animal containment system. The costs include regular inspections of rearing units, solids storage units, and drug/pesticide storage units. EPA considers the aquatic animal containment system to include any physical barriers and practices used to prevent the release of materials from the containment system. For flow-through and recirculating facilities, the containment system includes wastewater treatment, for example, quiescent zones or settling basins, in addition to the rearing units and storage units. For net pens, the containment system includes the use of double nets or other techniques that may be used to deter predators. EPA also included costs for reporting of structural failure or damage to the containment system that results in a material discharge of pollutants to waters of the U.S.

For net pen systems, failures include physical damage to the predator control nets or the nets containing the aquatic animals, which result in a discharge of the contents of the nets. Damage includes abrasion, cutting or tearing of the nets and breakdown of the netting due to rot or ultra violet exposure. For flow-through and recirculating systems,

a failure includes a collapse or damage of a rearing unit or wastewater treatment structure; damage to pipes, valves, and other plumbing fixtures; and damage or malfunction to screens or physical barriers in the system, which would prevent the unit from containing water, sediment, and the aquatic animals. The rule provides the permitting authorities may specify what constitutes damage and/or a material discharge on a site-specific basis for the purposes of triggering the reporting requirement. Based on available information related to containment system failures in the past, flow-through and recirculating facilities have had less incidences of failures than net pen facilities. Therefore, EPA estimated that 10 percent of the flow-through and recirculating facilities would incur a cost associated with the reporting of the failure whereas, for costing purposes, all net pen facilities were assumed to experience a failure. Again, EPA believes these assumptions are conservative and may overestimate the frequency of reportable failures.

EPA revised estimates for all labor costs using the employee and wage information supplied in the detailed surveys. For those facilities indicating they use unpaid labor for all or part of the facility operation, or that did not supply useable wage information, EPA used average State or regional wages for both staff and management labor. Separate estimates were used for commercial and non-commercial facilities.

2. What Are the Total National Costs?

Tables IX–1 and IX–2 summarize numbers of affected facilities and total annualized costs for today's final

regulation. EPA estimates that a total of 242 facilities will be affected by today's final regulation. These counts include two non-profit flow-through facilities in Alaska producing 100,000 lb/year or more that did not receive a detailed questionnaire. More information is provided in the rulemaking record (DCN 63065). Table IX–1 summarizes the estimated number and type of facilities affected by the rule, based on the production threshold of 100,000 lb/year. These 242 facilities consists of 101 commercial facilities and 141 noncommercial facilities; noncommercial facilities include Federal, state, Alaskan non-profit, and Tribal hatcheries. Of the 101 commercial facilities, 32 are projected to be unprofitable prior to the final rule (i.e., baseline closures) under cash flow analysis. EPA did not identify any academic/research facilities in the detailed questionnaire that produced 100,000 lbs/yr or more.

The estimated cost for this rule is $1.4 million per year (pre-tax, 2003 dollars). Noncommercial facilities account for about 81 percent of the total cost of the rule. These estimated total costs reflect aggregate compliance costs incurred by facilities that produce 100,000 lb/year or more and will be affected by today's final regulation. EPA's total cost estimates do not include costs that are incurred by the 32 commercial facilities that are considered baseline closures. To the extent that some projected baseline closures remain open and incur costs under this rule, despite analysis showing unprofitability in the baseline, national compliance costs, pollutant load reductions and potential benefits would be higher than projected.

TABLE IX–1.—ESTIMATED NUMBER OF AFFECTED FACILITIES WITH PRODUCTION 100,000 LBS/YR OR MORE

Organization	Estimated number of facilities (see note)		
	Baseline closures [1]	Not baseline closures [2]	Total
Commercial	32 (28)	69[4] (69)	101 (97)
Noncommercial [3]	NA (NA)	141 (141)	141 (141)
Total	32 (28)	210 (210)	242 (238)

Note: Numbers in (parentheses) are facilities that are determined not to be in compliance with final rule requirements at the time this final rule is signed by the EPA Administrator.

NA: EPA does not determine closures for noncommercial facilities.

[1] Projected baseline closures are estimated using cash flow analysis. When net income analysis is assumed for earnings, the number of commercial baseline closures increases to 43. Baseline closures would not be projected to incur costs for a new rule in accordance with EPA's Guidelines for Preparing Economic Analyses (USEPA, EPA 240–R–00–003). Baseline closures (based on cash flow) are therefore not included in estimates of costs for this rule.

[2] Total costs and economic impacts for this rule are estimated using incremental compliance costs incurred by the facilities that are not baseline closures and not in compliance with the rule at time of final signature (i.e., 210 facilities are expected to incur costs under this rule: 69 commercial and 141 noncommercial facilities).

[3] Noncommercial facilities include those operated by States, Tribes, the Federal Government, and Alaskan Non-Profits.

[4] Includes two facilities that are projected to be baseline closures using discounted cash flow analysis but are characterized by EPA as "Not Baseline Closures" due to unique facility-specific evidence associated with production, fish type, scale, and financial data (as outlined in DCN 20500 in the confidential record for this rule).

TABLE IX–2.—NATIONAL COSTS: TOTAL BY SUBCATEGORY

Production system	Owner	Pre-tax annualized costs ($000, 2003 dollars)
		Final option
Flow-through and Recirculating Systems	Commercial	$256
	Noncommercial [2]	$1,149
Net Pen	Commercial	$36
	Noncommercial [2]	$0
Total pre-tax [1]	....	$1,442

Note: Totals may not sum due to rounding.
[1] Total annual post-tax cost for the final option is $1,362.
[2] Noncommercial facilities include those operated by State, Federal, Alaska nonprofit, and Tribal facilities.

B. Economic Impacts

This section discusses the economic effects associated with the final rule.

1. How did EPA Estimate Economic Effects?

Existing Commercial Facilities. EPA uses several measures to evaluate possible impacts on existing commercial facilities. These measures examine the possibility of business closure and corresponding direct impacts on employment and communities and indirect and national impacts associated with closures. EPA also evaluates potential moderate impacts short of closure, as well as changes in financial health and borrowing capacity.

To evaluate impacts to commercial facilities, EPA conducts a closure analysis that compares projected earnings, with and without cost of compliance with the final regulation for the period 2005 to 2015. For this rule, EPA used discounted cash flow and net income to estimate earnings for closure analysis. The difference between cash flow and net income is depreciation (cash flow equals net income plus depreciation). Analysis using net income is more likely to identify baseline closures and could demonstrate additional regulatory closures associated with the rule. Table IX–3.5 presents closure results obtained using both discounted cash flow and net income. All other analytical results (for example, other measures of economic impacts, costs and benefits) presented in this final action reflect discounted cash flow as the basis for earnings. EPA also examines the effects of attributing a wage rate to unpaid labor and found that imputing costs for unpaid labor and management would not change the projected economic impacts of the rule.

Closure analysis assumes that (1) producers are unable to pass on the costs of incremental pollution control to consumer through higher prices and (2) costs and earnings are discounted

assuming a 7 percent real discount rate to account for the time value of money and place earnings and costs on a comparable basis. EPA considers that the rule will result in a facility closure if a facility shows (1) positive discounted cash flow (or net income) without the rule and (2) negative discounted cash flow (or net income) with the rule for two out of three forecasting scenarios. The forecasting methods give a range of trends: (1) Optimistic or upward (USDA CPI Food at Home, Fish and Seafood Sector), (2) pessimistic or downward (weighted average, based on facility production, of USDA trout price data or U.S. Department of Labor, Bureau of Labor Statistics, Fish PPI, Producer Price Index—Unprocessed and packaged fish, not seasonally adjusted), and (3) neutral or no change (average of 1999–2001 earnings collected in the detailed questionnaire). In an effort to evaluate the effects of relying on two out of three forecasts to define closures, EPA also analyzed closures using a more conservative assumption whereby closures are defined as occurring when negative earnings are projected under only one of three forecast scenarios.

EPA does not assess potential for closure under the rule if a facility is projected to have negative earnings under baseline conditions (*i.e.*, baseline closure). Baseline closures are defined as facilities that are projected to have negative earnings under 2 or 3 of the forecasting methods before they incur pollution control costs (*i.e.*, baseline closures). EPA's standard methodology when using forecasts in closure models is to use a "weight of evidence" approach across a set of reasonable assumptions regarding future industry behavior. This allows EPA to recognize uncertainty in the forecasts without placing undue emphasis on any one set of "timing and initial conditions". Using this methodology, EPA determined that 32 out of 101

commercial facilities are baseline closures, assuming discounted cash flow for earnings. When EPA adopts net income as the basis for earnings, baseline closures are projected to be 43. When EPA projects closures based on negative earnings in one out of three forecasts, baseline closures are projected to be 34. EPA notes that this type of analysis identifies candidates for closure; information on facility-level costs and earnings may be too uncertain to allow precise prediction of which operations will actually close, in the absence of the rule.

In addition to its closure analysis, EPA also prepared additional analyses to assess potential effects, short of closure, on existing businesses, including an analysis of additional moderate impacts using a sales test, an evaluation of financial health using an approach similar to that used by USDA, and an assessment of possible impacts on borrowing capacity. Use of these measures has the advantage that they mirror analyses that investment and lending institutions perform to evaluate industries and businesses.

First, to assess whether there are additional moderate impacts to facilities, EPA uses a sales test to compare the pre-tax annualized cost of the final rule to the revenues reported for facilities that passed the baseline closure analysis. EPA considers that facilities show additional moderate impacts if they are not projected to close but incur compliance costs in excess of 5 percent of facility revenue; this threshold is consistent with threshold values established by EPA in previous regulations and is determined to be appropriate for this rulemaking.

Second, EPA calculates impacts on financial health at the company level using USDA's 2×2 matrix (*i.e.*, four-level) categorization of financial health based on a combination of net cash income and debt/asset ratios. The categories are favorable, marginal solvency, marginal income, and vulnerable. EPA considers any change in financial health category as an impact of the rule.

Finally, EPA performs a credit test by calculating the ratio of the pre-tax annualized cost of an option and the after-tax Maximum Feasible Loan Payment (MFLP) (*i.e.*, 80 percent of after-tax cash flow). EPA identified companies with a ratio exceeding 80 percent of MFLP as being impacted by this rule (*i.e.*, the test threshold is therefore actually 64 percent of the after-tax cash flow).

For the purposes of EPA's analysis, the Agency assumes (1) no growth in production to offset incremental costs and (2) that the costs of the rule are not passed on to consumers. The facility must absorb all increased costs. If it cannot do so and remain in operation, all production is assumed lost. EPA's assumption of no cost pass through is a conservative approach to evaluating economic achievability among regulated entities. To evaluate market and trade level impacts, EPA assumes all costs are shifted onto the broader market level as a way of assessing the upper bound of potential impacts.

The Economic and Environmental Benefit Analysis, available in the rulemaking record, provides more detail on EPA's analysis (DCN 63010).

Noncommercial Facilities. For today's final rule, EPA collected information on how U.S. Fish and Wildlife Service and State agencies make decisions about operating or closing public hatcheries. EPA confirmed that public hatcheries close; the U.S. Fish and Wildlife Service hatchery system once had as many as 250 hatcheries and it now operates fewer than 90 facilities. Closures may result from funding cuts (*e.g.*, Mitchell Act Funds and the Willard National Fish Hatchery or General Funds for State Hatcheries) or revision of a program's mission and goals (*e.g.*, increase focus on endangered species versus provision of recreational services). Closures may also result from water quality impacts associated with aquaculture activities. The costs of upgrading pollution control at public hatcheries are not generally the primary reason for closure, but costs may tip the balance of a particular hatchery toward a closure decision. See the Economic and Environmental Benefits Analysis (DCN 63010) for more details.

In the absence of well defined tests for projecting public facility closures, EPA compares pre-tax annualized compliance costs to 2001 operating budgets for public facilities ("Budget Test"). For the purposes of this analysis, costs exceeding 5 percent and 10

percent are assumed to signal potential "moderate" and "adverse" impacts, respectively. EPA examines the ability of State-owned hatcheries to recoup compliance costs through increases in funding derived solely from user fees. All States and the District of Columbia have fishing license fees for residents. The license fees are not raised every year even though costs increase through inflation. Instead, when fees are raised or a fish stamp instituted, the incremental or new fee is usually a round number such as $3, $5, or $10. A $3 to $5 hike in State fishing license fees translates into an increase in fees of about 20 percent to 35 percent. Although all States report having fishing license fees, if a state hatchery reports no funding from user fee sources, EPA considers that facility to be unable to recoup increased costs through increased funding from user fees.

More detailed information is provided in the Economic and Environmental Benefit Analysis and the rulemaking record.

New Commercial Facilities. To assess effects on new businesses, EPA's analysis considers the barrier that compliance costs due to the effluent guidelines regulation may pose to entry into the industry. In general, it is less costly to incorporate waste water treatment technologies as a facility is built than it is to retrofit existing facilities. Therefore, where a rule is economically achievable for existing facilities, it will also be economically achievable for new facilities that can meet the same guidelines at lower cost. Similarly, even where the cost of compliance with a given technology is not economically achievable for an existing source, such technology may be less costly for new sources and thus have economically sustainable costs. It is possible, on the other hand, that to the extent the up-front costs of building a new facility are significantly increased as a result of the rule, prospective builders may face difficulties in raising additional capital. This could present a barrier to entry. Therefore, as part of its analysis of new source standards, EPA evaluates barriers to entry. If the requirements promulgated in the final regulation do not give existing operators a cost advantage over new source operators, then EPA assumes new source performance standards do not present a barrier to entry for new facilities.

EPA's analysis includes all commercial facilities within scope of the rule, including those that are baseline closures. EPA examines the (1) proportion of commercial facilities that incur no costs, (2) proportion of

commercial facilities that incur no land or capital costs, and (3) ratio of incremental land and capital costs to total company assets. The cost to asset ratio is calculated using company data because asset data were collected only at the company level; company impacts cannot be extrapolated to the national-level because sampling weights are based on facilities, not companies. EPA calculates the ratio for each company and uses the average of the ratios. More information is provided in the Economic and Environmental Impact Analysis available in the rulemaking record.

2. What Are the Results of the Economic Analysis?

Existing Commercial Facilities. Table IX–3 shows the impacts on commercial operations from today's regulation. As shown, EPA projects no facility closures as a result of the final rule under the cash flow analysis. No closures are projected for enterprises or companies. Correspondingly, there are no employment and other direct and indirect impacts estimated for this rule as a consequence of closures using cash flow analysis and negative earnings in two of three forecast scenarios. When the closure analysis is conducted using net income as a basis for earnings, EPA projects two closures out of 58 commercial facilities (*see* Table IX–3.5). When the closure analysis is conducted using only one of three forecast scenarios, EPA also identifies two closures out of 67 commercial facilities (*see* Section IX.B.1 for discussion of forecast methods). Based on these results, EPA concludes that the final rule option is economically achievable. EPA notes that all other analytical results (for example other measures of economic impacts, costs) presented in this final action reflect discounted cash flow as the basis for earnings; EPA's analyses indicate that use of net income will not materially change results.

EPA expects some operations will incur moderate impacts, short of closure, based on an analysis that shows that some operations will incur compliance costs in excess of 5 percent of annual revenue. For the final regulation, 4 of 69 commercial facilities incur costs greater than 5 percent of sales, affecting about 5 percent of regulated facilities in the flow-through and recirculating subcategory; no additional facilities have costs exceeding 3 percent of revenues. No commercial facilities have costs that exceed 10 percent of annual revenue. EPA's analysis shows no expected change in financial health. One company fails the USDA credit test as a result of the final regulation. These results are based on data from companies represented in the Agency's detailed questionnaire. These results further support EPA's conclusion that the final options are economically achievable for commercial facilities (and companies). More information is provided in the Economic and Environmental Benefit Analysis available in the rulemaking record (DCN 63010)

Noncommercial Facilities. Table IX–3 also shows the impacts on noncommercial operations from today's regulation. Four facilities incur costs exceeding 10 percent of budget. EPA assumes that those facilities that face costs exceeding 10 percent of their budget would be adversely affected by the final regulation. None of these facilities report the use of user fee funds. These results indicate that 3 percent of all non-commercial operations may be adversely affected by the final option. Under EPA's assumed criteria for determining economic achievability, these operations may be vulnerable to closure.

Twelve facilities incur costs exceeding 5 percent of annual budgets under the final rule. These results indicate that an additional 6 percent of all non-commercial operations (not counting those adversely affected) would experience some moderate impact, short of closure, associated under this final rule. Some of these facilities report the use of user fees revenues, implying potential flexibility in meeting the incremental costs.

No in-scope Alaskan nonprofit facilities responded to EPA's detailed questionnaire, but EPA did identify two in-scope facilities based on screener data. These facilities were costed using screener data and economic impacts were projected based on publicly available revenue data for 2001. Neither facility is projected to incur costs greater than 3 percent of revenues.

Given that the results of EPA's analysis project that a small share of regulated noncommercial facilities may incur costs exceeding 10 percent of budget, estimated at 3 percent of facilities, the Agency has determined that these final technology options to be economically achievable for noncommercial facilities. For more information, see the Economic and Environmental Benefit Analysis available in the rulemaking record.

New Commercial Facilities. EPA estimated that about 4 percent of regulated facilities do not incur any costs under the final regulation, and about 76 percent of facilities incur no land or capital costs. The incremental land and capital costs, where they were incurred, represented less than 0.2 percent of total assets. This final regulation should therefore not present barriers to entry for new businesses.

TABLE IX–3.—ECONOMIC IMPACTS: EXISTING COMMERCIAL & NONCOMMERCIAL OPERATIONS

Threshold test	Number of in-scope facilities in the Analysis [1]	Impacts projected under final option
Commercial Operations		
Closure Analysis (discounted cash flow) [2]	69	0
Sales test >3% (facility level)	69	4
Sales test >5% (facility level)	69	4
Sales test >10% (facility level)	69	0
Change in Financial Health (Company level) [3]	34	0
Credit test >80% (Company level) [3]	34	1
Noncommercial Facilities [6]		
Budget test >3% (all facilities)	141	19
State owned only (# with user fees) [5]	106	12 (8)
Federal owned only	33	7
Alaskan Non-Profit [4]	2	0
Budget test >5% (all facilities)	141	12
State owned only (# with user fees) [5]	106	8 (8)
Federal owned only	33	4
Alaskan Non-Profit [4]	2	0
Budget test >10% (all facilities)	141	4
State owned only (# with user fees) [5]	106	0 (0)
Federal owned only	33	4
Alaskan Non-Profit [4]	2	0

Source: Estimated by USEPA using results from facility-specific detailed questionnaire responses, see Chapter 3.

[1] There are 101 in-scope commercial facilities, represented by 34 unweighted companies. Of the 101 facilities, 32 are baseline closures, assuming cash flow analysis, leaving 69 commercial facilities that can be analyzed. Closure analysis and sales test are performed at facility level; financial health and credit tests performed at company level; and all noncommercial tests performed at facility level.

[2] Closure analysis results obtained using discounted cash flow and closure defined as negative earnings in two of three forecast scenarios. *See* Table IX–3.5 for results under different assumptions.

[3] Analysis performed at the company level. The statistical weights, however, are developed on the basis of facility characteristics and therefore cannot be used for estimating the number of companies.

[4] Two Alaska non-profit organizations are within the scope of this rule, but did not receive a detailed survey. They were costed using screener survey data. Economic impacts were calculated using publically available information.

[5] Some State-owned facilities reported that they relied, in part, on funds from State user fee operations. These numbers are reported in parenthesis and are included in the overall numbers as well.

[6] There is a potential for a small number of Tribal facilities to be present within the population of non-commercial facilities, despite the absence of a line item for Tribal facilities above. In its screener survey which was a census of the industry, EPA identified a number of Tribal facilities that might be subject to the proposed rule for the CAAP category (DCN 51401). However, all of the tribal facilities represented by the detailed survey were determined to not be in scope.

Because the detailed survey is a sample, there is uncertainty associated with the conclusion that there are no tribal facilities in scope for the final rule. For this reason, EPA believes there may be a few in-scope tribal facilities that have not been analyzed. As part of the analyses conducted prior to the NODA, based on the screener data, EPA estimated impacts for tribal facilities producing between 20,000 and 100,000 pounds per year for Option B (more costly than the final option). These results are for facilities that are not within the scope of the final rule, but they provide evidence that the final rule is expected to be economically achievable for tribal facilities.

TABLE IX–3.5.—CLOSURE ANALYSIS FOR COMMERCIAL FACILITIES UNDER DIFFERENT ASSUMPTIONS

	Number of in-scope facilities in the analysis [1]	Closures projected under final option
Closure Analysis (discounted cash flow) [2]	69	0
Closure Analysis (Net Income) [2]	58	2
Closure Analysis (one out of three forecasts) [3]	67	2

[1] There are 32, 43, and 34 baseline closures projected under discounted cash flow, net income and one out of three forecasts respectively. Baseline closures are not analyzed for regulatory closure and therefore subtracted from the 101 in-scope facilities.

[2] Discounted cash flow and net income are two different assumptions used to estimate earnings under closure analysis (*see* Section IX.B.1 for details). Closures defined as occurring when negative earnings are projected under at least two of three forecast methods.

[3] Analysis assumes earnings estimated using cash flow and closure defined, more conservatively, as occurring when negative earnings are projected under only one of three forecast methods.

3. What Are the Projected Market Level Impacts?

EPA was not able to prepare a market model analysis for this rule because of the complex interaction between commercial and non-commercial operations (*e.g.*, trout are raised commercially, but also for restoration and recreation), wild catch accounts for a large share of the market for some species, and USDA Census data indicate that there is a high degree of concentration of specific species, such as trout and some other food fish. Literature on estimated measures of elasticity of supply and demand is limited and exist for only a few species, such as catfish which are not covered by this regulation. The Agency does therefore not report quantitative estimates of changes in overall supply and demand for aquaculture products and changes in market prices. For more information, *see* Chapter 3.6 of the Economic and Environmental Benefit Analysis for the proposed rulemaking available in the docket (DCN 63010). However, EPA does not expect significant market impacts as a result of today's final rule because economic impacts are expected to be low (*see* discussion above) and the overall cost of the rule is low, as compared to the total value of the U.S. aquaculture industry. Long-term shifts in supply associated with this rule are unlikely given expected continued competition from domestic wild harvesters and low-cost foreign suppliers. For additional information, see the Economic and Environmental Impact Analysis available in the rulemaking record.

4. What Are the Potential Impacts on Foreign Trade?

Foreign trade impacts are difficult to predict, since agricultural exports are determined by economic conditions in foreign markets and changes in the international exchange rate for the U.S. dollar. In addition, for today's final rule, EPA was not able to perform a market model analysis for this rule and did not obtain quantitative estimates of changes in overall supply and demand for aquaculture products and changes in market prices, as well as changes in traded volumes including imports and exports.

Nevertheless, EPA believes that the impact of this final rule on U.S. aquaculture trade will not be significant. Because of the relatively small market share of U.S. aquaculture producers in world markets, EPA believes that long-term shifts in supply associated with this rule are unlikely given expected continued competition from domestic wild harvesters and already lower-cost foreign suppliers in China and other Asian nations. Under a scenario that assumes the total costs of the rule are absorbed by the domestic market, EPA estimates that U.S. aquaculture prices would rise by slightly more than 1 cent per pound. Under the alternative assumption that all costs are born by facility operators, impacts are projected to be small and would not significantly affect production (*see* Section IX.B.2).

5. What Are the Potential Impacts on Communities?

The communities where aquaculture facilities are located may be affected by the final regulation if facilities cut back operations. However, EPA projects no commercial facility closures as a result of this rule, assuming discounted cash flow (two closures are projected using net income as shown in Table IX–3.5), indicating minimal likelihood of measurable impacts on (1) direct losses in commercial production, revenue, or employment; and (2) local economies and employment rates. Should some facilities cut back operations as a result of this final regulation, EPA cannot project how great these impacts would be as it cannot identify the communities where impacts might occur. Under a scenario that assumes the total costs of the rule are absorbed by the domestic market, EPA estimates that U.S. aquaculture prices would rise by slightly more than 1 cent per pound. (See EPA's Economic and Environmental Benefit Analysis.)

Closures of non-commercial facilities could also result in employment impacts on communities. EPA projects four noncommercial facilities, with a total employment of 16 employees could experience impacts such that they would be vulnerable to closure (*i.e.*, costs exceed 10 percent of annual budget). The communities in which these facilities are located could experience moderate impacts, but, as noted in Section IX.B.2, environmental compliance costs are generally a contributing rather than the deciding factor in closure decisions. EPA therefore does not expect significant impacts on communities as a result of today's final rule.

C. What Do the Cost-Reasonableness Analyses Show?

EPA performed an assessment of the total cost of the final rule relative to the expected effluent reductions. EPA based its "cost reasonableness" (CR) analysis on estimated costs, loadings, and removals. See EPA's Development Document in the rulemaking record for additional details.

Table IX.4 shows the cost-reasonableness values for conventional pollutants. EPA estimates BOD and TSS removals for each facility for each option. Because BOD can be correlated with TSS, EPA selected the higher of the two values (not the sum) to avoid possible double-counting of removals. For the Flow-through and Recirculating Systems Subcategory, cost-reasonableness is $2.77/lb. Cost-reasonableness is undefined for the Net Pen Subcategory systems because these facilities have adequate treatment to achieve requirements for pollutants (*i.e.,* no incremental removals are estimated for these facilities).

TABLE IX–4.—COST-REASONABLENESS: BOD OR TSS

Subcategory	Pre-tax annualized costs ($2003)	BOD or TSS removals (lb) [1]	Cost-reasonableness ($2003/pound)
Flow-through and Recirculating Systems	$1,405,866	506,839	$2.77
Net pen	$35,640	0	Undefined

[1] EPA determines the higher of BOD or TSS mass removal for each facility and then aggregates pounds across facilities.

Undefined: Facilities in this group are not projected to achieve incremental removals of the pollutants in this table (*i.e.,* no incremental removals are estimated).

X. What Are the Environmental Benefits for This Rule?

A. Summary of Environmental Benefits

Today's final action does not establish numeric limits for total suspended solids (TSS) or other pollutants from flow-through and recirculating systems. It establishes BMPs for solids control, materials storage, structural maintenance, recordkeeping, and training. The final rule also requires the permittee to develop a BMP plan on-site describing how the permittee will achieve the BMP requirements and make the plan available to the permitting authority upon request. The facilities are also to maintain the structural integrity of the aquatic animal containment system. The final rule also establishes BMP requirements for net pen systems that address feed management, waste collection and disposal, discharges associated with transport and harvest, carcass removal, materials storage, structural maintenance, recordkeeping, and training. Net pen facilities are to develop and maintain a BMP plan on-site describing how the permittee is to achieve the BMP requirements. The permittee must make the plan available to the permitting authority upon request. Both the flow-through and recirculating and net pen subcategories have reporting requirements for (1) the use of INADs and extralabel drugs use, (2) failure or damage to the structural integrity of the aquatic animal containment system, and (3) spills of drugs, pesticides and feed which result in discharge of pollutants to waters of the U.S. The requirements, according to EPA loadings estimates, will reduce facility discharges of TSS, total nitrogen (TN), total phosphorus (TP), and biochemical oxygen demand (BOD). EPA has also estimated reductions for metals and some feed contaminants as a result of these final requirements. EPA could not quantify baseline or regulated loads for drugs and pesticides.

These requirements and loading reductions (TSS, TN, TP, BOD, metals, and feed contaminants) could affect water quality, the uses supported by varying levels of water quality, and other aquatic environmental variables (*e.g.,* primary production and populations or assemblages of native organisms in the receiving waters of regulated facilities). These impacts may result in environmental benefits, some of which have quantifiable, monetizable value to society. For today's final action, EPA has only monetized benefits from water quality improvements resulting from reductions in TSS, TN, TP, and BOD.

TABLE 1.—SUMMARY OF ENVIRONMENTAL BENEFITS OF FINAL RULE

Type of benefit	Monetized value ($2003)
Improved water quality from reduced TSS, TN, TP, and BOD loadings due to improved solids control, including feed management	$66,000–$99,000
Reduced inputs to receiving water of metals and feed contaminants	not monetized
Reduced inputs of drugs and pesticides	not monetized
Reduced inputs of materials as a result of structural maintenance and material storage requirements	not monetized

B. Non-Monetized Benefits

1. Metals and Other Additives and Contaminants

CAAP facilities may release metals and other feed additives and contaminants to the environment in limited quantities; proper management of solids and other management practices may reduce environmental risk from these releases. Trace amounts of metals are added to feed in the form of mineral packs to ensure that the essential dietary nutrients are provided. In general, FDA establishes safety limits for feed additives and must address environmental safety concerns associated with such additives under the requirements of the Federal Food, Drug, and Cosmetic Act (FFD&CA) and National Environmental Policy Act (NEPA). Trace amounts of metals may also be present as feed contaminants. Metals may also be introduced into the environment from CAAP machinery, equipment, and structures (*e.g.,* net pens treated with antifouling copper compounds). Other feed additives may include FDA-approved compounds used to improve the coloring of fish flesh. Organochlorine contaminants such as polychlorinated biphenyls (PCBs) also may be present as trace residues regulated by FDA in some fish feeds.

EPA estimates that today's final rule will reduce total suspended solids (TSS) released by CAAP facilities by about half a million pounds per year. Metals and other feed contaminants that may be released to the environment from CAAP facilities are in large part associated with waste solids. EPA estimates that reductions in TSS will be accompanied by incidental removals of metals and PCBs. EPA estimated metal reductions of approximately 2,700 pounds per year nationally and a maximum of PCB reductions of 0.04 lbs per year. For further discussion of metals and other feed additives and contaminants, see the Economic and Environmental Impact Analysis and Technical Development Document for this final rule (DCNs 63010 and 63009).

2. Drugs and Pesticides

CAAP facilities employ drugs and pesticides for a variety of therapeutic and water treatment purposes. Facilities release treated waters that may contain residual amounts of drugs, pesticides, and their byproducts directly to the environment. Drugs used for therapeutic purposes are regulated by FDA. Prior to approving drugs for use, FDA must evaluate the environmental safety of animal drugs as required by FFDCA and NEPA. While FDA is required to consider environmental impacts of approved and investigational drugs under these authorities, the environmental safety of drugs used under FDA's "investigational new animal drug" (INAD) program may not be fully characterized. The INAD program is an important mechanism that enables the collection of data that can be used to characterize and establish the environmental safety of new drugs. For compilations of technical literature supporting FDA's environmental assessments of therapeutants used at CAAP facilities, see the FDA's Center for Veterinary Medicine (CVM) Web site (*www.fda.gov/cvm*). It should be noted that FDA environmental assessments are not site-specific and may not cover all discharge scenarios (*e.g.*, multiple dischargers to a single receiving water) or applications (*e.g.*, extralabel applications of drugs). For additional discussion of this topic, see Chapter 7 of EPA's Environmental Impact Analysis for this final rule.

Today's final rule requires the proper storage of drugs, pesticides, and feed to prevent spills that may result in a discharge from CAAP facilities. For reasons explained in Section VI.G (Loadings) of this Preamble, EPA has not quantified expected reductions in the release of drugs and pesticides to the environment nor environmental benefits that might result. Today's final rule also requires CAAP facilities to report to permitting authorities whenever an investigative drug or an extralabel drug is used in amounts exceeding a previously approved dosage, as described above in Section VIII.E. This requirement is expected to better enable permitting authorities to monitor the potential for environmental risks that could result from such uses. EPA has not quantified benefits that might arise as a result of this requirement.

C. Monetized Benefits

1. Case Study Framework

As was done for EPA's proposed rule, EPA estimated monetized benefits of the regulation based on predicted improvements in water quality in the receiving waters of facilities that were expected to have load reductions as a result of the rule. EPA's water quality modeling for today's final action differs from the proposal modeling, however, in that for the final rule, more detailed, facility-specific operational and environmental data were obtained, both from information provided by facilities on the detailed surveys as well as other sources. This more detailed data provided EPA with a better basis for developing representative case studies on which to perform water quality modeling and valuation and for extrapolating from case studies to a national benefit estimate.

To select a set of representative case studies from among the facilities for which EPA had detailed data, EPA assumed that three factors primarily drive water quality improvements at any given facility: (1) The magnitude of pollutant load reductions under the final rule, (2) effluent pollutant concentrations at baseline (prior to regulatory reductions), and (3) the ratio of facility effluent flow to receiving water streamflow ("dilution ratio"). EPA then created categories based on combinations of values (low and high) for each of these factors. For example, the "LLL" category means facilities with "low" pollutant reductions under the final rule, "low" baseline effluent concentrations, and "low" dilution ratios; this category is expected to experience the smallest benefits of the final regulation. In this manner, eight categories were created (LLL, LLH, LHL, LHH, HLL, HLH, HHL, HHH; *see* Table 2). EPA then assigned all detailed survey facilities with non-zero load reductions in the scope of the final rule to an appropriate category based on the three factors described above. For more details on the categorization procedure, *see* Chapter 8 of the Economic and Environmental Impact Analysis for today's final action [DCN 63010].

EPA then developed a "case study" for one facility in each of the five categories expected to experience the greatest water quality improvement (EPA did not develop case studies for all categories partly because of resource constraints). EPA multiplied the estimated benefits for each case study by the total number of facilities assigned to that category to estimate a total national benefit for that category. No benefits were estimated for the three

categories for which case studies were not developed. Benefits for these categories are expected to be small relative to those included in the analysis. The total national benefit estimate was estimated as the sum of benefits for all categories.

2. Economic Valuation Method

Economic research indicates that the public is willing to pay for improvements in water quality and several methods have been developed to translate changes in water quality to monetized values, as noted in EPA's "Guidelines for Preparing Economic Analyses (EPA–240–R–00–003, 2003;). At proposal, EPA based the water quality benefits monetization on results from a stated-preference survey conducted by Carson and Mitchell (1993) (DCN 20157). We divided household willingness-to-pay (WTP) values for changes in recreational water "use classes" by the number of "water quality index" points (an index based on water quality variables; see below) in each use class. We assigned a portion of the value for each unit change to achieving the whole step. Recently, EPA developed an alternative approach, also based on Mitchell and Carson's work. Mitchell and Carson also expressed their results as an equation relating a household's WTP for improved water quality to the change in the water quality index and household income. An important feature of this approach is that it is less sensitive to the baseline use of the water body. This approach is also consistent with economic theory in that it exhibits a declining marginal WTP for water quality (see more information on this approach in DCNS 40138 and 40595). While caution must be used in manipulating valuations derived from stated preference surveys, this valuation function approach helps address some concerns about earlier applications of the water quality benefits monetization method. (*See* DCN 40595 for a more detailed discussion).

3. Water Quality Modeling

As was done for the proposed rule, EPA applied the Enhanced Stream Water Quality Model (QUAL2E, *http://www.epa.gov/waterscience/wqm/*) to simulate changes in receiving water quality resulting from reductions in TSS, BOD, total nitrogen, and total phosphorus estimated by EPA to result from the regulatory requirements of this final rule. QUAL2E is a one-dimensional water quality model that assumes steady state flow but allows simulation of diurnal variations in temperature, algal photosynthesis, and respiration. The model projects water quality by solving an advective-dispersive mass transport equation. Water quality constituents simulated include conservative substances, temperature, bacteria, BOD_5, DO, ammonia, nitrate and organic nitrogen, phosphate and organic phosphorus, and algae.

Resource and data limitations constrained the number of QUAL2E applications that could be performed. EPA developed a QUAL2E case study for the following categories: LHL, LHH, HLH, HHL, and HHH. EPA did not prepare case studies for the LLL, LLH, and HLL categories because (a) no facilities were in the HLL category and (b) EPA focused modeling resources on categories expected to represent a larger proportion of benefits. Water quality improvements for facilities in the LLL and LLH categories were expected to be smaller than the improvements for the facilities in the other categories.

4. Calculation of "Water Quality Index"

Simulated water quality changes for each case study must be translated into a composite "index" value for the monetization method described in Section X.B.2 above. EPA more recently developed a six-parameter WQI ("WQI–6") based on TSS, BOD, DO, FC, plus nitrate (NO_3) and phosphate (PO_4). The new index more completely reflects the type of water quality changes that will result from loading reductions for TSS, total nitrogen (TN), total phosphorus (TP), and BOD. Final rule benefits presented here were estimated on the basis of WQI–6.

5. Estimated National Water Quality Benefits

EPA monetized water quality benefits for each of the 5 QUAL2E case studies performed (Table 2). Using the methods described above, the Agency estimates that the total national benefit from water quality improvements arising from TSS, BOD, TN, and TP reductions from this rule are $66,000—$99,000. This range reflects varying assumptions that the Agency implemented to reflect some sources of uncertainty. Furthermore, this range of water quality-based benefits of this regulation may be uncertain for several reasons including:

• EPA did not estimate benefits for the facilities in the LLL and LLH extrapolation categories. However, it is not expected that inclusion of these facilities would greatly increase monetized water quality benefits.

• EPA's monetization method mainly captures benefits for recreational uses of the streams. Economic research indicates that there are significant "non-use" values associated with some

dimensions of water quality. Analysis using monetization methods that fully captures non-use values could increase the estimated benefits for this rule if it significantly affects these dimensions. EPA does not have enough information to determine if this is the case.

• Other receiving water impacts are not captured in the QUAL2E modeling, such as build-up of organic sediments in stream channels. Research included in the administrative record for today's final action documents that such accumulations can impair aquatic ecosystems. Benefits from reducing these effects are not captured in EPA's analysis of water quality-based benefits of today's final action.

TABLE 2.—EXTRAPOLATED TOTAL NATIONAL WATER QUALITY BENEFIT ESTIMATE, FINAL OPTION

A Extrapolation category	B Total national benefit for extrapolation category ($2003)
LLL–LLH	not estimated
LHL–LHH	$2,126–$5,330
HLL–HLH	$6,591–$12,031
HHL–HHH	$57,497–$81,255
Total	$66,214–$98,616

In general, however, the relatively small recreational benefits projected for the rule suggest that non-monetized benefits categories are likely to be small as well.

XI. What Are the Non-Water Quality Environmental Impacts of This Rule?

Under Sections 304(b) and 306 of the Clean Water Act, EPA may consider non-water quality environmental impacts (including energy requirements) when developing effluent limitations guidelines and standards. Accordingly, EPA has considered the potential impact of today's final regulation on air emissions, energy consumption, and solid waste generation.

A. Air Emissions

With the implementation of feed management, the final rule decreases the amount of solid waste generated and land applied from CAAP facilities. Land application is a common waste disposal method in the CAAP industry; therefore, the amount of ammonia released as air emissions would be expected to decrease as the quantity of waste applied to cropland decreases. EPA estimates the decrease in ammonia emissions to be 8,182 pounds of ammonia per year. This is a decrease of about 8 % over the ammonia emissions

presently estimated for the industry. For additional details about air emissions from CAAP facilities, see Chapter 11 of the TDD.

B. Energy Consumption

EPA estimates that implementation of today's rule would result in a net decrease in energy consumption for aquaculture facilities. The decrease would be based on electricity used today to pump solids from raceways to solids settling ponds, which will no longer be generated, from wastewater treatment equipment. EPA determined that the decrease in energy consumption for flow-through and recirculating systems is estimated at 4,900 kilowatt-hour (kW-h). This represents about 1.3 $\times 10^{-7}$ percent of the national generated energy.

C. Solid Waste Generation

EPA estimates that implementation of today's rule would result in an estimated reduction of 2.3 million pounds of sludge, on a wet basis (assuming 12 percent solids) for flow-through and recirculating facilities. This reduction is due to feed management that results in less solid waste generated.

XII. How Will This Rule Be Implemented?

This section helps permit writers and CAAP facilities implement this regulation. This section also discusses the relationship of upset and bypass provisions, variances, and modifications to the final limitations and standards. For additional implementation information, see Chapter 2 of the Technical Development Document for today's rule.

A. Implementation of Limitations and Standards for Direct Dischargers

Effluent limitations guidelines and new source performance standards act as important mechanisms to control the discharges of pollutants to waters of the United States. These limitations and standards are applied to individual facilities through NPDES permits issued by the EPA or authorized States under Section 402 of the Act.

In specific cases, the NPDES permitting authority may elect to establish technology-based permit limits for pollutants not covered by this regulation. In addition, where State water quality standards or other provisions of State or Federal law require limits on pollutants not covered by this regulation (or require more stringent limits or standards on covered pollutants in order to attain and maintain water quality standards), the permitting authority must apply those limitations or standards. See CWA Section 301(b)(1)(C).

The final regulation establishing narrative limitations for the flow-through and recirculating system and net pen subcategories requires that a point source must meet the prescribed limitations expressed as operational practices or "any modification to these requirements as determined by the permitting authority based on its exercise of its best professional judgment." Sections 451.11 and 451.21. This provision authorizes the permitting authority to tailor the specific NPDES permit limits that implement the guideline limitations to individual sites. As previously explained, the final narrative requirements, in many cases, require achievement of environmental end points. There may be circumstances which require some modification to these requirements to best accomplish these environmental end points, or to accommodate specific circumstances at a particular site. The provision allows the permitting authority to address such situations by incorporating in the NPDES permit specific tailored conditions that accomplish the intent of the narrative limitations. The CWA recognizes that it should provide mechanisms for addressing certain unique, site-specific situations in the guidelines regulation. Here, EPA has provided upfront in this rule such a mechanism.

1. What Are the Compliance Dates for Existing and New Sources?

New and reissued NPDES permits to direct dischargers must include these effluent limitations unless water quality considerations require more stringent limits, and the permits must require immediate compliance with such limitations. If the permitting authority wishes to provide a compliance schedule, it must do so through an enforcement mechanism.

New sources must comply with the new source standards (NSPS) of this rule when they commence discharging CAAP wastewater. Because the final rule was not promulgated within 120 days of the proposed rule, the Agency considers a discharger to be a new source if its construction commences after September 22, 2004.

2. Who Does Part 451 Apply To?

In Section VI.A. of this preamble and Chapter 2 of the TDD, EPA provides detailed information on the applicability of this rule. 40 CFR part 451 will apply to existing and new concentrated aquatic animal production facilities that produce 100,000 pounds

or more of aquatic animals per year in flow-through, recirculating, and net pen systems. There is an exception for net pen systems rearing native species released after a growing period of no longer than 4 months to supplement commercial and sport fisheries.

B. Upset and Bypass Provisions

A "bypass" is an intentional diversion of the streams from any portion of a treatment facility. An "upset" is an exceptional incident in which there is unintentional and temporary noncompliance with technology-based permit effluent limitations because of factors beyond the reasonable control of the permittee. EPA's regulations concerning bypasses and upsets for direct dischargers are set forth at 40 CFR 122.41(m) and (n) and for indirect dischargers at 40 CFR 403.16 and 403.17.

C. Variances and Modifications

While the CWA requires application of effluent limitations established pursuant to section 301 to all direct dischargers, the statute also provides for the modification of these national requirements in a limited number of circumstances. Moreover, the Agency established administrative mechanisms to provide an opportunity for relief from the application of the national effluent limitations guidelines for categories of existing sources for toxic, conventional, and nonconventional pollutants.

1. Fundamentally Different Factors Variances

EPA will develop effluent limitations or standards different from the otherwise applicable requirements if an individual discharging facility is fundamentally different with respect to factors considered in establishing the limitation of standards applicable to the individual facility. Such a modification is known as a "fundamentally different factors" (FDF) variance.

Early on, EPA, by regulation provided for the FDF modifications from the BPT effluent limitations, BAT limitations for toxic and nonconventional pollutants and BCT limitations for conventional pollutants for direct dischargers. FDF variances for toxic pollutants were challenged judicially and ultimately sustained by the Supreme Court. (*Chemical Manufacturers Assn* v. *NRDC*, 479 U.S. 116 (1985)).

Subsequently, in the Water Quality Act of 1987, Congress added new Section 301(n) of the Act explicitly to authorize modifications of the otherwise applicable BAT effluent limitations or categorical pretreatment standards for existing sources if a facility is

fundamentally different with respect to the factors specified in Section 304 (other than costs) from those considered by EPA in establishing the effluent limitations or pretreatment standard. Section 301(n) also defined the conditions under which EPA may establish alternative requirements. Under Section 301(n), an application for approval of a FDF variance must be based solely on (1) information submitted during rulemaking raising the factors that are fundamentally different or (2) information the applicant did not have an opportunity to submit. The alternate limitation or standard must be no less stringent than justified by the difference and must not result in markedly more adverse non-water quality environmental impacts than the national limitation or standard.

EPA regulations at 40 CFR Part 125, Subpart D, authorizing the Regional Administrators to establish alternative limitations and standards, further detail the substantive criteria used to evaluate FDF variance requests for direct dischargers. Thus, 40 CFR 125.31(d) identifies six factors (*e.g.*, volume of process wastewater, age and size of a discharger's facility) that may be considered in determining if a facility is fundamentally different. The Agency must determine whether, on the basis of one or more of these factors, the facility in question is fundamentally different from the facilities and factors considered by EPA in developing the nationally applicable effluent guidelines. The regulation also lists four other factors (*e.g.*, infeasibility of installation within the time allowed or a discharger's ability to pay) that may not provide a basis for an FDF variance. In addition, under 40 CFR 125.31(b) (3), a request for limitations less stringent than the national limitation may be approved only if compliance with the national limitations would result in either (a) a removal cost wholly out of proportion to the removal cost considered during development of the national limitations, or (b) a non-water quality environmental impact (including energy requirements) fundamentally more adverse than the impact considered during development of the national limits.

The legislative history of Section 301(n) underscores the necessity for the FDF variance applicant to establish eligibility for the variance. EPA's regulations at 40 CFR 125.32(b)(1) are explicit in imposing this burden upon the applicant. The applicant must show that the factors relating to the discharge controlled by the applicant's permit which are claimed to be fundamentally different are, in fact, fundamentally different from those factors considered by EPA in establishing the applicable guidelines. In practice, very few FDF variances have been granted for past ELGs. An FDF variance is not available to a new source subject to NSPS or PSNS.

Facilities must submit all FDF variance applications to the appropriate Director (defined at 40 CFR 122.2) no later than 180 days from the date the limitations or standards are established or revised (see CWA section 301(n)(2) and 40 CFR 122.21(m)(1)(i)(B)(2)). EPA regulations clarify that effluent limitations guidelines are "established" or "revised" on the date those effluent limitations guidelines are published in the **Federal Register** (see 40 CFR 122.21 (m)(1)(i)(B)(2)). Therefore, all facilities requesting FDF variances from the effluent limitations guidelines in today's final rule must submit FDF variance applications to their Director (as defined at 40 CFR 122.2) no later than February 21, 2005.

2. Economic Variances

Section 301(c) of the CWA authorizes a variance from the otherwise applicable BAT effluent guidelines for nonconventional pollutants due to economic factors. The request for a variance from effluent limitations developed from BAT guidelines must normally be filed by the discharger during the public notice period for the draft permit. Other filing time periods may apply, as specified in 40 CFR 122.21(1)(2). Specific guidance for this type of variance is available from EPA's Office of Wastewater Management.

D. Best Management Practices

Sections 304(e), 308(a), 402(a), and 501(a) of the CWA authorize the Administrator to prescribe BMPs as part of effluent limitations guidelines and standards or as part of a permit. EPA's BMP regulations are found at 40 CFR 122.44(k). Section 304(e) of the CWA authorizes EPA to include BMPs in effluent limitations guidelines for certain toxic or hazardous pollutants for the purpose of controlling "plant site runoff, spillage or leaks, sludge or waste disposal, and drainage from raw material storage." Section 402(a)(1) and NPDES regulations [40 CFR 122.44(k)] also provide for best management practices to control or abate the discharge of pollutants when numeric limitations and standards are infeasible. In addition, Section 402(a)(2), read in concert with Section 501(a), authorizes EPA to prescribe as wide a range of permit conditions as the Administrator deems appropriate in order to ensure compliance with applicable effluent

limitations and standards and such other requirements as the Administrator deems appropriate.

E. Potential Tools To Assist With the Remediation of Aquaculture Effluents

A potential option to assist land owners with aquaculture effluent quality is the Environmental Quality Incentives Program (EQIP). This is a voluntary USDA conservation program. EQIP was reauthorized in the Farm Security and Rural Investment Act of 2002 (Farm Bill 2002). The Natural Resources Conservation Service (NRCS) administers EQIP funds.

EQIP applications are accepted throughout the year. NRCS evaluates each application using a state and locally developed evaluation process. Incentive payments may be made to encourage a producer to adopt land management, manure management, integrated pest management, irrigation water management and wildlife habitat management practices or to develop a Comprehensive Nutrient Management Plan (CNMP). These practices would provide beneficial effects on reducing sediment and nutrient loads to those aquaculture operations dependent on surface water flows. In addition, opportunities exist to provide EQIP funds to foster the adoption of innovative cost effective approaches to address a broad base of conservation needs, including aquaculture effluent remediation. NRCS does not at present have standards that apply specifically to waste handling at aquaculture facilities, thus EQIP funds for aquaculture projects would only apply to practices related to other agricultural aspects of a facility such as CNMPs for the land application of solids.

XIII. Statutory and Executive Order Reviews

A. Executive Order 12866: Regulatory Planning and Review

Under Executive Order 12866, [58 FR 51,735 (October 4, 1993)] the Agency must determine whether the regulatory action is "significant" and therefore subject to OMB review and the requirements of the Executive Order. The Order defines "significant regulatory action" as one that is likely to result in a rule that may:

(1) Have an annual effect on the economy of $100 million or more or adversely affect in a material way the economy, a sector of the economy, productivity, competition, jobs, the environment, public health or safety, or State, local or tribal governments or communities;

(2) Create a serious inconsistency or otherwise interfere with an action taken or planned by another agency;

(3) Materially alter the budgetary impact of entitlements, grants, user fees, or loan programs or the rights and obligations of recipients thereof; or

(4) Raise novel legal or policy issues arising out of legal mandates, the President's priorities, or the principles set forth in the Executive Order.

Pursuant to the terms of Executive Order 12866, it has been determined that this rule is a "significant regulatory action." As such, this action was submitted to OMB for review. Changes made in response to OMB suggestions or recommendations will be documented in the public record.

B. Paperwork Reduction Act

The information collection requirements in this rule have been submitted for approval to the Office of Management and Budget (OMB) under the *Paperwork Reduction Act,* 44 U.S.C. 3501 *et seq.* The information collection requirements are not enforceable until OMB approves them.

EPA has several special reporting and monitoring provisions in this regulation as previously explained. The provisions include reporting requirements (1) for the use of INAD or extralabel drug uses; (2) for failure or damage to the containment system (including the production system(s) and all the associated storage and water treatment systems) that results in a material discharge of pollutants to waters of the U.S; and (3) for spills of drugs, pesticides or feed. Section 308(a) of the CWA authorizes the Administrator to require the owner or operator of any point source to file reports as required to carry out the objectives of the Act. This ELG requires reporting in the event that drugs are used which are either under a conditional approval as an Investigative New Animal Drugs (INADs) or are prescribed by a licensed veterinarian for treatment of a disease or a species that is outside the approved use of the specific drug, referred to as extralabel drug use, unless the INAD or extralabel drug use is under similar conditions and dosages as a previously approved use. EPA believes this reporting requirement is appropriate for these classes of drugs, because they have not undergone the same degree of review with respect to their environmental effects as approved drugs. The final regulation also requires reporting when the facility has a failure in the structural integrity of the aquatic animal containment systems that results in a material discharge of pollutants. EPA believes this reporting is necessary to alert the permitting authority to the release of large quantities of material from these facilities. The rule also allows the permitting authority to specify in the permit what constitutes damage and/or material discharge of pollutants for particular facilities based on consideration of relevant site-specific factors.

Burden means the total time, effort, or financial resources expended by persons to generate, maintain retain, or disclose or provide information to or for a Federal agency. This includes the time needed to review instructions; develop, acquire, install, and utilize technology and systems for the purposes of collecting, validating, and verifying information, processing and maintaining information, and disclosing and providing information; search data sources; complete and review the collection of information; and transmit or otherwise disclose the information. EPA estimates that the reporting and recordkeeping requirements included in today's regulation will result in a total annual burden of 45,000 hours and cost $808,000.

An agency may not conduct or sponsor, and a person is not required to respond to a collection of information unless it displays a currently valid OMB control number. The OMB control numbers for EPA's regulations in 40 CFR are listed in 40 CFR part 9. When this ICR is approved by OMB, the Agency will publish a technical amendment to 40 CFR part 9 in the **Federal Register** to display the OMB control number for the approved information collection requirements contained in this final rule.

C. Regulatory Flexibility Act

The RFA generally requires an agency to prepare a regulatory flexibility analysis of any rule subject to notice and comment rulemaking requirements under the Administrative Procedure Act or any other statute unless the agency certifies that the rule will not have a significant economic impact on a substantial number of small entities. Small entities include small businesses, small organizations, and small governmental jurisdictions.

For purposes of assessing the impacts of today's rule on small entities, small entity is defined as: (1) A small business that is primarily engaged in concentrated aquatic animal production, as defined by North American Industry Classification (NAIC) codes 112511 and 112519, with no more than $0.75 million in annual revenues; (2) a small governmental jurisdiction that is a government of a city, county, town, school district or special district with a

population of less than 50,000; and (3) a small organization that is any not-for-profit enterprise which is independently owned and operated and is not dominant in its field.

After considering the economic impacts of today's final rule on small entities, I certify that this action will not have a significant economic impact on a substantial number of small entities. The small entities directly regulated by the final rule are primarily commercial businesses that fall within the NAIC codes for finfish farming, fish hatcheries, and other aquaculture. The Small Business Administration size standard for these codes is $0.75 million in annual revenues. Among the costed facilities, EPA identified 38 facilities belonging to small businesses or organizations. Of the 38, 37 facilities are owned by small businesses and 1 is an Alaskan facility operated by a small non-profit organization that is not dominant in its field. For the purposes of the RFA, Federal, and State governments are not considered small governmental jurisdictions, as documented in the rulemaking record (DCN 20121). Thus, facilities owned by these governments are not considered small entities, regardless of their production levels. EPA identified no public facilities owned by small local governments. No small organization is projected to incur impacts. Of the 101 commercial facilities, 37 (37 percent) are owned by small businesses. Under EPA's closure analyses no small business is projected to close as a result of the final rule, assuming discounted cash flow (two small business closures are projected using net income). In addition to considering the potential for adverse economic impacts, EPA also evaluated the possibility of other, more moderate financial impacts. Expressed as a comparison of compliance costs to sales, only 4 facilities belonging to small businesses (11 percent of small businesses, and 4 percent of commercial facilities) are likely to incur costs that exceed 3 percent of sales. One small business fails the USDA credit test.

Although this final rule will not have a significant economic impact on a substantial number of small entities, EPA nonetheless designed the rule to reduce the impact on small entities. The scope of the final rule is restricted to CAAP facilities that produce 100,000 lbs/year or more. This means that of the approximately 4,000 aquaculture facilities nationwide, as identified by USDA's Census of Aquaculture, EPA's final regulation applies to an estimated 101 commercial facilities or approximately 2.6 percent of all operations. Among commercial

facilities, EPA identifies 38 facilities (37 percent of in-scope facilities) as small businesses using SBA's definition. Finally, EPA based the final rule on a technology option that has lower costs and fewer impacts (including impacts on small businesses) than several other technology options that were considered as possible bases for the final rule.

EPA conducted outreach to small entities and convened a Small Business Advocacy Review Panel prior to proposal to obtain the advice and recommendations of representatives of the small entities that potentially would be subject to the rule's requirements. The Agency convened the Small Business Advocacy Review Panel on January 22, 2002. Members of the Panel represented the Office of Management and Budget, the Small Business Administration, and EPA. The Panel met with small entity representatives (SERs) to discuss the potential effluent guidelines and, in addition to the oral comments from SERs, the Panel solicited written input. In the months preceding the Panel, EPA conducted outreach with small entities that would potentially be affected by this regulation. On January 25, 2002, the SBAR Panel sent some initial information for the SERs to review and provide comment on. On February 6, 2002, the Panel distributed additional information to the SERs for their review. On February 12 and 13, the Panel met with SERs to hear their comments on the information distributed in these mailings. The Panel also received written comments from the SERs in response to the discussions at this meeting and the outreach materials. The Panel asked SERs to evaluate how they would be affected and to provide advice and recommendations regarding early ideas to provide flexibility. See Section 8 of the Panel's Report (DCN 31019) for a complete discussion of SER comments. The Panel evaluated the assembled materials and small-entity comments on issues related to the elements of an Initial Regulatory Flexibility Analysis. A copy of the Panel's report is included in the rulemaking docket. EPA provided responses to the Panel's most significant findings in the Notice of Proposal Rulemaking (67 FR 57918–57920). In general, the requirements of this final rule address the concerns raised by SERs and are consistent with the Panel's recommendations.

D. Unfunded Mandates Reform Act

Title II of the Unfunded Mandates Reform Act of 1995 (UMRA), Public Law 104–4, establishes requirements for Federal agencies to assess the effects of their regulatory actions on State, local, and tribal governments and the private sector. Under section 202 of the UMRA, EPA generally must prepare a written statement, including a cost-benefit analysis, for proposed and final rules with "Federal mandates" that may result in expenditures to State, local, and tribal governments, in the aggregate, or to the private sector, of $100 million or more in any one year. Before promulgating an EPA rule for which a written statement is needed, section 205 of the UMRA generally requires EPA to identify and consider a reasonable number of regulatory alternatives and adopt the least costly, most cost-effective or least burdensome alternative that achieves the objectives of the rule. The provisions of section 205 do not apply when they are inconsistent with applicable law. Moreover, section 205 allows EPA to adopt an alternative other than the least costly, most cost-effective or least burdensome alternative if the Administrator publishes with the final rule an explanation why that alternative was not adopted. Before EPA establishes any regulatory requirements that may significantly or uniquely affect small governments, including tribal governments, it must have developed under section 203 of the UMRA a small government agency plan. The plan must provide for notifying potentially affected small governments, enabling officials of affected small governments to have meaningful and timely input in the development of EPA regulatory proposals with significant Federal intergovernmental mandates, and informing, educating, and advising small governments on compliance with the regulatory requirements.

EPA has determined that this rule does not contain a Federal mandate that may result in expenditures of $100 million or more for State, local, and tribal governments, in the aggregate, or the private sector in any one year. The total annual cost of this rule is estimated to be $1.4 million. Thus, today's rule is not subject to the requirements of Sections 202 and 205 of UMRA.

E. Executive Order 13132: Federalism

Executive Order 13132, entitled "Federalism" (64 FR 43255, August 10, 1999), requires EPA to develop an accountable process to ensure "meaningful and timely input by State and local officials in the development of regulatory policies that have federalism implications." "Policies that have federalism implications" is defined in the Executive Order to include regulations that have "substantial direct effects on the States, on the relationship between the national government and

the States, or on the distribution of power and responsibilities among the various levels of government."

This rule does not have Federalism implications. It will not have substantial direct effects on the States, on the relationship between the national government and the States, or on the distribution of power and responsibilities among the various levels of government, as specified in Executive Order 13132. EPA estimates that, when promulgated, these revised effluent guidelines and standards will be incorporated into NPDES permits without significant additional costs to authorized States.

Further, the revised regulations would not alter the basic State-Federal scheme established in the Clean Water Act under which EPA authorizes States to carry out the NPDES permitting program. EPA expects the revised regulations to have little effect, if any, on the relationship between, or the distribution of power and responsibilities among, the Federal, State and local governments. Thus, Executive Order 13132 does not apply to this rule.

F. Executive Order 13175: Consultation and Coordination With Indian Tribal Governments

Executive Order 13175, entitled "Consultation and Coordination with Indian Tribal Governments" (65 FR 67249, November 9, 2000), requires EPA to develop an accountable process to ensure "meaningful and timely input by tribal officials in the development of regulatory policies that have tribal implications." "Policies that have tribal implications" is defined in the Executive Order to include regulations that have substantial direct effects on one or more Indian tribes, on the relationship between the Federal government and the Indian tribes, or on this distribution of power and responsibilities between the Federal government and Indian tribes."

The final rule does not have tribal implications. It will not have substantial direct effects on tribal governments, on the relationship between the Federal government and Indian tribes, or on the distribution of power and responsibilities between the Federal government and Indian tribes, as specified in Executive Order 13175. The Executive Order provides that EPA must ensure meaningful and timely input by tribal officials in the development of regulatory policies that have tribal implications. EPA's rulemaking process has provided that opportunity for meaningful and timely input. EPA first published a notice of proposed

rulemaking for CAAPs in September 2002, requesting comment on the proposal. In December 2003, EPA issued a Notice of Data Availability describing options for changes to the proposed rule. As noted, EPA identified a number of tribal facilities in its screener survey, however further evaluation did not identify any in-scope tribal facilities based on subsequent evaluation of the detailed survey information from a sample of these facilities. Thus EPA has not had a basis to have any formal consultation with Tribal officials. EPA has however concluded that the final rule will not have a substantial direct effect on one or more Indian Tribes, will not impose substantial direct compliance costs on Indian tribal governments, nor pre-empt tribal law.

G. Executive Order 13045: Protection of Children From Environmental Health and Safety Risks

Executive Order 13045 (62 FR 19885, April 23, 1997) applies to any rule that: (1) Is determined to be "economically significant" as defined under Executive Order 12866, and (2) concerns an environmental health or safety risk that EPA has reason to believe may have a disproportionate effect on children. If the regulatory action meets both criteria, the Agency must evaluate the environmental health and safety effects of the planned rule on children, and explain why the planned regulation is preferable to other potentially effective and reasonably feasible alternatives considered by the Agency.

This rule is not subject to Executive Order 13045 because it is not an economically significant rule under E.O. 12866.

H. Executive Order 13211: Actions That Significantly Affect Energy Supply, Distribution, or Use

This rule is not a "significant energy action" as defined in Executive Order 13211, "actions concerning Regulations that Significantly Affect Energy Supply, Distribution, or Use" (66 FR 28355 (May 22, 2001)) because it is not likely to have a significant adverse effect on the supply, distribution, or use of energy. As part of the Agency's consideration of non-water quality impacts, EPA has estimated the energy consumption associated with today's requirements. The rule will result in a net decrease in energy consumption for flow-through and recirculating systems. The decrease would be based on electricity used today to pump solids from raceways to solids settling ponds, which will no longer be generated, from wastewater treatment equipment. EPA estimated the decrease in energy consumption for flow-through and recirculating systems at 4,900 kilowatt-hour (kW-h). Comparing the annual decrease in electric use resulting from the final requirements to national annual energy use, EPA estimates the decrease to be 1.3×10^{-7} percent of national energy use. Therefore, we conclude that this rule is not likely to have any adverse energy effects.

I. National Technology Transfer and Advancement Act

As noted in the proposed rule, Section 12(d) of the National Technology Transfer and Advancement Act of 1995 ("NTTAA"), Public Law 104–113, 12(d) (15 U.S.C. 272 note) directs EPA to use voluntary consensus standards in its regulatory activities unless to do so would be inconsistent with applicable law or otherwise impractical. Voluntary consensus standards are technical standards (*e.g.*, materials specifications, test methods, sampling procedures, and business practices) that are developed or adopted by voluntary consensus standards bodies. The NTTAA directs EPA to provide Congress, through OMB, explanations when the Agency decides not to use available and applicable voluntary consensus standards. Today's rule does not establish any technical standards, thus NTTAA does not apply to this rule.

J. Executive Order 12898: Federal Actions To Address Environmental Justice in Minority Populations and Low-Income Populations

The requirements of the Environmental Justice Executive Order are that EPA will review the environmental effects of major Federal actions significantly affecting the quality of the human environment. For such actions, EPA reviewers will focus on the spatial distribution of human health, social and economic effects to ensure that agency decision makers are aware of the extent to which those impacts fall disproportionately on covered communities. This is not a major action. Further, EPA does not believe this rulemaking will have a disproportionate effect on minority or low income communities because the technology-based effluent limitations guidelines are uniformly applied nationally irrespective of geographic location. The final regulation will reduce the negative effects of concentrated aquatic animal production industry waste in our nation's waters to benefit all of society, including minority and low-income communities. The cost impacts of the rule should likewise not disproportionately affect low-income

communities given the relatively low economic impacts of today's final rule.

K. Congressional Review Act

The Congressional Review Act, 5 U.S.C. 801 *et seq.*, as added by the Small Business Regulatory Enforcement Fairness Act of 1996, generally provides that before a rule may take effect, the agency promulgating the rule must submit a rule report, which includes a copy of the rule, to each House of the Congress and to the Comptroller General of the United States. EPA will submit a report containing this rule and other required information to the U.S. Senate, the U.S. House of Representatives, and the Comptroller General of the United States prior to publication of the rule in the **Federal Register**. A major rule cannot take effect until 60 days after it is published in the **Federal Register**. This action is not a "major rule" as defined by 5 U.S.C. 804(2). This rule will be effective September 22, 2004.

List of Subjects in 40 CFR Part 451

Environmental protection, Concentrated aquatic animal production, Waste treatment and disposal, Water pollution control.

Dated: June 30, 2004.

Stephen L. Johnson,

Acting Deputy Administrator.

■ For the reasons set forth in the preamble, chapter I of title 40 of the Code of Federal Regulations is amended by adding part 451 to read as follows:

PART 451—CONCENTRATED AQUATIC ANIMAL PRODUCTION POINT SOURCE CATEGORY

Sec.
451.1   General applicability.
451.2   General definitions.
451.3   General reporting requirements.

Subpart A—Flow-Through and Recirculating Systems Subcategory

451.10   Applicability.
451.11   Effluent limitations attainable by the application of the best practicable control technology currently available (BPT).
451.12   Effluent limitations attainable by the application of the best available technology economically achievable (BAT).
451.13   Effluent limitations attainable by the application of the best conventional technology (BCT).
451.14   New source performance standards (NSPS).

Subpart B—Net Pen Subcategory

451.20   Applicability.
451.21   Effluent limitations attainable by the application of the best practicable control technology currently available (BPT).

451.22   Effluent limitations attainable by the application of the best available technology economically achievable (BAT).
451.23   Effluent limitations attainable by the application of the best conventional technology (BCT).
451.24   New source performance standards (NSPS).

Authority: 7 U.S.C. 135 *et seq.*, 136–136y; 15 U.S.C. 2001, 2003, 2005, 2006, 2601–2671, 21 U.S.C. 331j, 346a, 348; 31 U.S.C. 9701; 33 U.S.C. 1251 *et seq.*, 1311, 1313d, 1314, 1318, 1321, 1326, 1330, 1342, 1344, 1345(d) and (e), 1361; 42 U.S.C. 241, 242b, 243, 246, 300f, 300g, 300g–1, 300g–2, 300g–3, 300g–4, 300g–5, 300g–6, 300j–2, 300j–3, 300j–4, 300j–9, 1857 *et seq.*, 6901–6992k, 7401–7671q, 7542, 9601–9657, 11023, 11048; E.O. 11735, 38 FR 21243, 3 CFR, 1971–1975 Comp., 973.

§ 451.1   General applicability.

As defined more specifically in each subpart, this Part applies to discharges from concentrated aquatic animal production facilities as defined at 40 CFR 122.24 and Appendix C of 40 CFR Part 122. This Part applies to the discharges of pollutants from facilities that produce 100,000 pounds or more of aquatic animals per year in a flow-through, recirculating, net pen or submerged cage system.

§ 451.2   General definitions.

As used in this part:
(a) The general definitions and abbreviations in 40 CFR part 401 apply.
(b) *Approved dosage* means the dose of a drug that has been found to be safe and effective under the conditions of a new animal drug application.
(c) *Aquatic animal containment system* means a culture or rearing unit such as a raceway, pond, tank, net or other structure used to contain, hold or produce aquatic animals. The containment system includes structures designed to hold sediments and other materials that are part of a wastewater treatment system.
(d) *Concentrated aquatic animal production facility* is defined at 40 CFR 122.24 and Appendix C of 40 CFR Part 122.
(e) *Drug* means any substance defined as a drug in section 201(g)(1) of the Federal Food, Drug and Cosmetic Act (21 U.S.C. 321).
(f) *Extralabel drug use* means a drug approved under the Federal Food, Drug and Cosmetic Act that is not used in accordance with the approved label directions, see 21 CFR part 530.
(g) *Flow-through system* means a system designed to provide a continuous water flow to waters of the United States through chambers used to produce aquatic animals. Flow-through systems typically use rearing units that are either raceways or tank systems.

Rearing units referred to as raceways are typically long, rectangular chambers at or below grade, constructed of earth, concrete, plastic, or metal to which water is supplied by nearby rivers or springs. Rearing units comprised of tank systems use circular or rectangular tanks and are similarly supplied with water to raise aquatic animals. The term does not include net pens.

(h) *Investigational new animal drug (INAD)* means a drug for which there is a valid exemption in effect under section 512(j) of the Federal Food, Drug, and Cosmetic Act, 21 U.S.C. 360b(j), to conduct experiments.

(i) *New animal drug application* is defined in 512(b)(1) of the Federal Food, Drug, and Cosmetic Act (21 U.S.C 360b(b)(1)).

(j) *Net pen system* means a stationary, suspended or floating system of nets, screens, or cages in open waters of the United States. Net pen systems typically are located along a shore or pier or may be anchored and floating offshore. Net pens and submerged cages rely on tides and currents to provide a continual supply of high-quality water to the animals in production.

(k) *Permitting authority* means EPA or the State agency authorized to administer the National Pollutant Discharge Elimination System permitting program for the receiving waters into which a facility subject to this Part discharges.

(l) *Pesticide* means any substance defined as a "pesticide" in section 2(u) of the Federal Insecticide, Fungicide, and Rodenticide Act (7 U.S.C. 136(u)).

(m) *Real-time feed monitoring* means a system designed to track the rate of feed consumption and to detect uneaten feed passing through the nets at a net pen facility. These systems may rely on a combination of visual observation and hardware, including, but not limited to, devices such as video cameras, digital scanning sonar, or upweller systems that allow facilities to determine when to cease feeding the aquatic animals. Visual observation alone from above the pens does not constitute real-time monitoring.

(n) *Recirculating system* means a system that filters and reuses water in which the aquatic animals are produced prior to discharge. Recirculating systems typically use tanks, biological or mechanical filtration, and mechanical support equipment to maintain high quality water to produce aquatic animals.

§ 451.3   General reporting requirements.

(a) *Drugs.* Except as noted below, a permittee subject to this Part must notify the permitting authority of the use in a concentrated aquatic animal production facility subject to this Part of any investigational new animal drug (INAD) or any extralabel drug use where such a use may lead to a discharge of the drug to waters of the U.S. Reporting is not required for an INAD or extralabel drug use that has been previously approved by FDA for a different species or disease if the INAD or extralabel use is at or below the approved dosage and involves similar conditions of use.

(1) The permittee must provide a written report to the permitting authority of an INAD's impending use within 7 days of agreeing or signing up to participate in an INAD study. The written report must identify the INAD to be used, method of use, the dosage, and the disease or condition the INAD is intended to treat.

(2) For INADs and extralabel drug uses, the permittee must provide an oral report to the permitting authority as soon as possible, preferably in advance of use, but no later than 7 days after initiating use of that drug. The oral report must identify the drugs used, method of application, and the reason for using that drug.

(3) For INADs and extralabel drug uses, the permittee must provide a written report to the permitting authority within 30 days after initiating use of that drug. The written report must identify the drug used and include: the reason for treatment, date(s) and time(s) of the addition (including duration), method of application; and the amount added.

(b) Failure in, or damage to, the structure of an aquatic animal containment system resulting in an unanticipated material discharge of pollutants to waters of the U.S. In accordance with the following procedures, any permittee subject to this Part must notify the permitting authority when there is a reportable failure.

(1) The permitting authority may specify in the permit what constitutes reportable damage and/or a material discharge of pollutants, based on a consideration of production system type, sensitivity of the receiving waters and other relevant factors.

(2) The permittee must provide an oral report within 24 hours of discovery of any reportable failure or damage that results in a material discharge of pollutants, describing the cause of the failure or damage in the containment system and identifying materials that have been released to the environment as a result of this failure.

(3) The permittee must provide a written report within 7 days of discovery of the failure or damage documenting the cause, the estimated time elapsed until the failure or damage was repaired, an estimate of the material released as a result of the failure or damage, and steps being taken to prevent a reccurrence.

(c) In the event a spill of drugs, pesticides or feed occurs that results in a discharge to waters of the U.S., the permittee must provide an oral report of the spill to the permitting authority within 24 hours of its occurrence and a written report within 7 days. The report shall include the identity and quantity of the material spilled.

(d) *Best management practices (BMP) plan.* The permittee subject to this Part must:

(1) Develop and maintain a plan on site describing how the permittee will achieve the requirements of § 451.11(a) through (e) or § 451.21(a) through (h), as applicable.

(2) Make the plan available to the permitting authority upon request.

(3) The permittee subject to this Part must certify in writing to the permitting authority that a BMP plan has been developed.

Subpart A—Flow-Through and Recirculating Systems Subcategory

§ 451.10   Applicability.

This subpart applies to the discharge of pollutants from a concentrated aquatic animal production facility that produces 100,000 pounds or more per year of aquatic animals in a flow-through or recirculating system.

§ 451.11   Effluent limitations attainable by the application of the best practicable control technology currently available (BPT).

Except as provided in 40 CFR 125.30 through 125.32, any existing point source subject to this subpart must meet the following requirements, expressed as practices (or any modification to these requirements as determined by the permitting authority based on its exercise of its best professional judgment) representing the application of BPT:

(a) *Solids control.* The permittee must:

(1) Employ efficient feed management and feeding strategies that limit feed input to the minimum amount reasonably necessary to achieve production goals and sustain targeted rates of aquatic animal growth in order to minimize potential discharges of uneaten feed and waste products to waters of the U.S.

(2) In order to minimize the discharge of accumulated solids from settling ponds and basins and production systems, identify and implement procedures for routine cleaning of

rearing units and off-line settling basins, and procedures to minimize any discharge of accumulated solids during the inventorying, grading and harvesting aquatic animals in the production system.

(3) Remove and dispose of aquatic animal mortalities properly on a regular basis to prevent discharge to waters of the U.S., except in cases where the permitting authority authorizes such discharge in order to benefit the aquatic environment.

(b) *Materials storage.* The permittee must:

(1) Ensure proper storage of drugs, pesticides, and feed in a manner designed to prevent spills that may result in the discharge of drugs, pesticides or feed to waters of the U.S.

(2) Implement procedures for properly containing, cleaning, and disposing of any spilled material.

(c) *Structural maintenance.* The permittee must:

(1) Inspect the production system and the wastewater treatment system on a routine basis in order to identify and promptly repair any damage.

(2) Conduct regular maintenance of the production system and the wastewater treatment system in order to ensure that they are properly functioning.

(d) *Recordkeeping.* The permittee must:

(1) In order to calculate representative feed conversion ratios, maintain records for aquatic animal rearing units documenting the feed amounts and estimates of the numbers and weight of aquatic animals.

(2) Keep records documenting the frequency of cleaning, inspections, maintenance and repairs.

(e) *Training.* The permittee must:

(1) In order to ensure the proper clean-up and disposal of spilled material adequately train all relevant facility personnel in spill prevention and how to respond in the event of a spill.

(2) Train staff on the proper operation and cleaning of production and wastewater treatment systems including training in feeding procedures and proper use of equipment.

§ 451.12 Effluent limitations attainable by the application of the best available technology economically achievable (BAT).

Except as provided in 40 CFR 125.30 through 125.32, any existing point source subject to this subpart must meet the following requirements representing the application of BAT: The limitations are the same as the corresponding limitations specified in § 451.11.

§ 451.13 Effluent limitations attainable by the application of the best conventional technology (BCT).

Except as provided in 40 CFR 125.30 through 125.32, any existing point source subject to this subpart must meet the following requirements representing the application of BCT: The limitations are the same as the corresponding limitations specified in § 451.11.

§ 451.14 New source performance standards (NSPS).

Any point source subject to this subpart that is a new source must meet the following requirements: The standards are the same as the corresponding limitations specified in § 451.11.

Subpart B—Net Pen Subcategory

§ 451.20 Applicability.

This subpart applies to the discharge of pollutants from a concentrated aquatic animal production facility that produces 100,000 pounds or more per year of aquatic animals in net pen or submerged cage systems, except for net pen facilities rearing native species released after a growing period of no longer than 4 months to supplement commercial and sport fisheries.

§ 451.21 Effluent limitations attainable by the application of the best practicable control technology currently available (BPT).

Except as provided in 40 CFR 125.30 through 125.32, any existing point source subject to this subpart must meet the following requirements, expressed as practices (or any modification to these requirements as determined by the permitting authority based on its exercise of its best professional judgment) representing the application of BPT:

(a) *Feed management.* Employ efficient feed management and feeding strategies that limit feed input to the minimum amount reasonably necessary to achieve production goals and sustain targeted rates of aquatic animal growth. These strategies must minimize the accumulation of uneaten food beneath the pens through the use of active feed monitoring and management practices. These practices may include one or more of the following: Use of real-time feed monitoring, including devices such as video cameras, digital scanning sonar, and upweller systems; monitoring of sediment quality beneath the pens; monitoring of benthic community quality beneath the pens; capture of waste feed and feces; or other good husbandry practices approved by the permitting authority.

(b) *Waste collection and disposal.* Collect, return to shore, and properly dispose of all feed bags, packaging materials, waste rope and netting.

(c) *Transport or harvest discharge.* Minimize any discharge associated with the transport or harvesting of aquatic animals including blood, viscera, aquatic animal carcasses, or transport water containing blood.

(d) *Carcass removal.* Remove and dispose of aquatic animal mortalities properly on a regular basis to prevent discharge to waters of the U.S.

(e) *Materials storage.*

(1) Ensure proper storage of drugs, pesticides and feed in a manner designed to prevent spills that may result in the discharge of drugs, pesticides or feed to waters of the U.S.

(2) Implement procedures for properly containing, cleaning, and disposing of any spilled material.

(f) *Maintenance.*

(1) Inspect the production system on a routine basis in order to identify and promptly repair any damage.

(2) Conduct regular maintenance of the production system in order to ensure that it is properly functioning.

(g) *Recordkeeping.*

(1) In order to calculate representative feed conversion ratios, maintain records for aquatic animal net pens documenting the feed amounts and estimates of the numbers and weight of aquatic animals.

(2) Keep records of the net changes, inspections and repairs.

(h) *Training.* The permittee must:

(1) In order to ensure the proper clean-up and disposal of spilled material adequately train all relevant facility personnel in spill prevention and how to respond in the event of a spill.

(2) Train staff on the proper operation and cleaning of production systems including training in feeding procedures and proper use of equipment.

§ 451.22 Effluent limitations attainable by the application of the best available technology economically achievable (BAT).

Except as provided in 40 CFR 125.30 through 125.32, any existing point source subject to this subpart must achieve the following effluent limitations representing the application of BAT: The limitations are the same as the limitations specified in § 451.21.

§ 451.23 Effluent limitations attainable by the application of the best conventional technology (BCT).

Except as provided in 40 CFR 125.30 through 125.32, any existing point source subject to this subpart must achieve the following effluent limitations representing the application of BCT: The limitations are the same as the limitations specified in § 451.21.

§ 451.24 New source performance standards (NSPS).

Any point source subject to this subpart that is a new source must meet the following requirements: The standard is the same as the limitations specified in § 451.21.

[FR Doc. 04–15530 Filed 8–20–04; 8:45 am]

BILLING CODE 6560–50–U

Appendix E1

BMP Plan Template

BMP Plan Template

You may want to use the following BMP plan template when writing your BMP plan. Fill in the sections marked in blue and/or italics.

Aquaculture Facility Name
Prepared: *Date*
NPDES Number: *# for your facility*
Facility Manager: *name, phone number*

A. Description of Facility

Provide a description of your facility. This description may include the following types of information:

- *Type of fish produced*
- *Annual amount of fish produced*
- *When the facility was constructed*
- *What type of systems (e.g., flow-through) are used at the facility*
- *Information about the systems (12 feet long raceways, etc.)*
- *Number of discharge points*

B. Water Source

Include a description of the source of the water at your facility. This description may include the following information:

- *Type of source – stream, ground, spring, etc.*
- *Name of the source (e.g., Upper Spring)*
- *If available, information about the quality of the water source (e.g., low in TSS)*
- *How the water arrives at the facility (e.g., ditch)*
- *Anything your facility does to treat incoming water (e.g., an inflow trash rack screen is used to catch vegetation from the spring and ditch prior to entering the facility. The trash rack screen is cleaned at least daily to prevent vegetation from affecting the water flow to facility)*

C. Treatment System(s) Used

Describe the treatment systems used at your facility. This description may include the following information:

- *Type of treatment system*
- *Design flow*
- *Normal operation*

- *Cleaning procedures*
- *Maintenance procedures*

D. Other Information

Provide any other additional information that might be useful to your permitting authority (e.g., additional information about how water flows into your facility or about oxygen recharge). In the following sections, describe in detail how you will achieve the specific requirements of the CAAP ELGs. Where helpful, you might attach example logs/forms used at your facility to physically show your permitting authority how you are complying with the CAAP ELGs.

E. Solids Control

FLOW-THROUGH AND/OR RECIRCULATING SYSTEMS

1. Efficient feed management (to limit feed input to the minimum amount reasonably necessary to achieve production goals and sustain targeted rates of aquatic animal growth).

   Describe the practices your facility uses to achieve efficient feed management. A form for tracking and calculating feed conversion ratios is available in Appendix N of the BMP Guidance.

2. Procedures for routine cleaning of rearing units and offline settling basins.

   Describe the cleaning procedures used. Also describe how your facility defines "routine." An example log to track cleaning is available in Appendix Q of the BMP Guidance.

3. Procedures for inventorying, grading, and harvesting aquatic animals (that minimize discharge of accumulated solids).

   Describe the procedures used.

4. Remove and dispose of aquatic animal mortalities properly on a regular basis to prevent discharge to waters of the United States (except where authorized by your permitting authority in order to benefit the aquatic environment).

   Describe the procedures for removal and disposal. A form for tracking carcass removal and disposal is available in Appendix T of the BMP Guidance.

F. Material Storage

FLOW-THROUGH, RECIRCULATING AND/OR NET PEN SYSTEMS

A form for tracking spills and leaks at your facility is available in Appendix O of the BMP Guidance.

1. Proper storage of drugs, pesticides, and feed to prevent spills that may result in the discharge to waters of the United States.

    Describe the practices used.

2. Procedures for properly containing, cleaning, and disposing of any spilled materials.

    Describe the procedures used.

G. Maintenance

Forms for tracking inspection and maintenance are available in Appendix P of the BMP Guidance.

FLOW-THROUGH AND/OR RECIRCULATING SYSTEMS

1. Routinely inspect production systems and wastewater treatment systems to identify and promptly repair damage.

    Describe the routine inspections performed. Also describe how your facility defines "routine."

2. Regularly conduct maintenance of production systems and wastewater treatment systems to ensure their proper function.

    Describe the regular maintenance performed. Also describe how your facility defines "regular."

NET PEN SYSTEMS

1. Routinely inspect production systems to identify and promptly repair damage.

    Describe the routine inspections performed. Also describe how your facility defines "routine."

2. Regularly conduct maintenance of production systems to ensure their proper function.

    Describe the regular maintenance performed. Also describe how your facility defines "regular."

H. Record-keeping

Use the checklist in Appendix R of the BMP Guidance to ensure that you are meeting the record-keeping requirements of the CAAP ELGs.

FLOW-THROUGH AND/OR RECIRCULATING SYSTEMS

1. Maintain records for aquatic animal rearing units documenting feed amounts and estimates of the numbers and weights of aquatic animals in order to calculate representative feed conversion ratios.

   Describe the records your facility keeps for documenting feed amounts and estimates of aquatic animals for calculating FCRs. A form for tracking and calculating FCRs is available in Appendix N of the BMP Guidance.

2. Keep records documenting frequency of cleaning, inspections, maintenance, and repairs.

   Describe the records your facility keeps to document this. Appendix P of the BMP Guidance contains forms for tracking inspection, maintenance, and repairs; Appendix Q of the BMP Guidance contains a form for tracking cleaning.

NET PEN SYSTEMS

1. Maintain records for aquatic animal rearing units documenting feed amounts and estimates of the numbers and weights of aquatic animals in order to calculate representative feed conversion ratios.

   Describe the records your facility keeps for documenting feed amounts and estimates of aquatic animals for calculating FCRs. A form for tracking and calculating FCRs is available in Appendix N of the BMP Guidance.

2. Keep records documenting net pen changes, inspections, and repairs.

   Describe the records your facility keeps to document this. Appendix P of the BMP Guidance contains forms for tracking inspection, maintenance, and repairs.

I. Training

Appendix S of the BMP Guidance contains a log for tracking employee training.

FLOW-THROUGH AND/OR RECIRCULATING SYSTEMS

1. Train all relevant personnel in spill prevention and how to respond in the event of a spill to ensure proper clean-up and disposal of spilled materials.

    Describe the procedures for training personnel in spill prevention and response.

2. Train personnel on proper operation and cleaning of production and wastewater treatment systems, including feeding procedures and proper use of equipment.

    Describe the procedures for training personnel on proper operation and cleaning.

NET PEN SYSTEMS

1. Train all relevant personnel in spill prevention and how to respond in the event of a spill to ensure proper clean-up and disposal of spilled materials.

    Describe the procedures for training personnel in spill prevention and response.

2. Train personnel on proper operation and cleaning of production systems, including feeding procedures and equipment.

    Describe the procedures for training personnel on proper operation and cleaning.

J. Feed Monitoring

NET PEN SYSTEMS

1. Employ efficient feed management and feeding strategies that limit feed input to the minimum amount reasonably necessary to achieve production goals and sustain targeted rates of aquatic animal growth.

   Describe the practices your facility uses to achieve efficient feed management. A form for tracking and calculating feed conversion ratios is available in Appendix N of the BMP Guidance.

2. Minimize accumulation of uneaten feed beneath the pens through active feed monitoring and management strategies approved by your permitting authority.

   Describe practices and management strategies to minimize uneaten feed beneath net pens.

K. Waste Collection and Disposal

NET PEN SYSTEMS NET

1. Collect, return to shore, and properly dispose of all feed bags, packaging materials, waste rope, and netting.

   Describe practices to accomplish this.

L. Transport or Harvest Discharge

NET PEN SYSTEMS

1. Minimize any discharge associated with the transport or harvesting of aquatic animals (including blood, viscera, aquatic animal carcasses, or transport water containing blood).

    Describe practices used to accomplish this.

M. Carcass Removal

NET PEN SYSTEMS ⬛

1. Remove and dispose of aquatic animal mortalities properly on a regular basis to prevent their discharge into waters of the United States.

   Describe procedures for removing and disposing of aquatic animal mortalities. Appendix T of the BMP Guidance contains a log for tracking carcass removal and disposal.

N. Diagram or Map

A diagram/map of the facility is helpful to illustrate the layout of the operation.

O. Review and Endorsement of the BMP Plan

We, the facility manager and the individuals responsible for implementing the BMP plan, have reviewed and endorsed this BMP plan.

_____ (Facility Name)	_____ (NPDES #)
_____ (Facility Manager – Printed Name)	_____ (Facility Manager – Signature)
_____ (Other Individual – Printed Name & Title)	_____ (Other Individual – Signature)
_____ (Other Individual – Printed Name & Title)	_____ (Other Individual – Signature)
_____ (Other Individual – Printed Name & Title)	_____ (Other Individual – Signature)

P. Certifying the BMP Plan with the Permitting Authority

Once your BMP plan has been developed and the facility manager and individuals responsible for implementing the BMP plan have reviewed and endorsed the plan, you must do the following:

1. Keep a copy of the BMP plan in your records. The plan must be made available to the permitting authority upon request.

2. Send a signed letter/form to your permitting authority stating that you have developed a BMP plan. The letter/form should include your name and title, name of the facility, NPDES number, and date the BMP plan was developed. An example certification form that may be submitted to your permitting authority is available in Appendix F of the BMP Guidance.

BMP Plan Checklist for Flow-Through and Recirculating Facilities

This checklist may be used to ensure that all required components are included in your BMP plan.

FACILITY DESCRIPTION

❑   A short description of your facility.

SOLIDS CONTROL

❑   Description of feed management/feeding strategies that limit feed input to achieve production goals and sustain targeted rates of aquatic animal growth, while minimizing potential discharges of uneaten feed/waste products to waters of the U.S.

❑   Description of procedures for routine* cleaning of rearing units and offline settling basins.

❑   Description of procedures for inventorying, grading, and harvesting aquatic animals that minimize discharge of accumulated solids.

❑   Description of the process for removing and disposing of aquatic animal mortalities on a regular basis to prevent discharge to waters of the United States, except where authorized by the permitting authority in order to benefit the aquatic environment.

MATERIAL STORAGE

❑   Description of procedures/practices to ensure proper storage of drugs, pesticides, and feed in a manner designed to prevent spills that may result in the discharge of drugs, pesticides, and feed to waters of the United States.

❑   Procedures for properly containing, cleaning, and disposing of any spilled materials.

STRUCTURAL MAINTENANCE

❑   Description of routine* procedures for inspecting production systems and wastewater treatment systems to identify and promptly repair damage.

❑   Description of regular* procedures for conducting maintenance of production systems and wastewater treatment systems to ensure their proper function.

RECORD-KEEPING

❑   Description of how you will maintain records for aquatic animal rearing units documenting feed amounts and estimates of the numbers and weights of aquatic animals to calculate FCRs.

❑   Description of how you will keep records documenting frequency of cleaning, inspections, maintenance, and repairs.

TRAINING

❑   Description of procedures for training all relevant personnel in spill prevention and how to respond to a spill to ensure proper clean-up and disposal of spilled materials.

❑   Description of procedures for training personnel on proper operation/cleaning of production and wastewater treatment systems (includes feeding procedures and proper equipment use).

CERTIFICATION

❑   Sent a letter to your permitting authority, certifying that a BMP Plan was developed for your facility. Refer to Appendix F for an example of a certification letter.

* Be sure to define "routine" and "regular" (which can vary during the year) in your BMP Plan.

BMP Plan Checklist for Net Pen Facilities

This checklist may be used to ensure all required components are included in your BMP plan.

FACILITY DESCRIPTION

❏ A short description of your facility.

FEED MANAGEMENT

❏ Description of feed management/feeding strategies that limit feed input to achieve production goals and sustain targeted rates of aquatic animal growth, while minimizing potential discharges of uneaten feed/waste products to waters of the U.S.

❏ Description of using active feed monitoring and management strategies (approved by the permitting authority) to minimize accumulation of uneaten feed beneath the pens.

WASTE COLLECTION AND DISPOSAL, TRANSPORT OR HARVEST DISCHARGE, CARCASS REMOVAL

❏ Description of how you will make sure to collect, return to shore, and properly dispose of all feed bags, packaging materials, waste rope, and netting.

❏ Description of practices to minimize discharge associated with transport or harvesting of aquatic animals (including blood, viscera, carcasses, or transport water containing blood).

❏ Description of procedures to ensure removal and disposal of aquatic animal mortalities properly on a regular basis to prevent their discharge into water of the U.S.

MATERIAL STORAGE

❏ Description of procedures/practices to ensure proper storage of drugs, pesticides, and feed to prevent spills that may result in discharge to waters of the U.S.

❏ Procedures for properly containing, cleaning, and disposing of any spilled materials.

MAINTENANCE

❏ Description of routine* procedures for inspecting production systems to identify/repair damage.

❏ Description of regular* procedures for conducting maintenance of production systems to ensure their proper function.

RECORD-KEEPING

❏ Description of how you will maintain records documenting feed amounts and estimates of numbers and weights of aquatic animals to calculate FCRs.

❏ Description of how you will document net changes, inspections, and repairs.

TRAINING

❏ Description of procedures for training all relevant personnel in spill prevention and how to respond to spills to ensure proper clean-up and disposal of spilled materials.

❏ Description of procedures for training personnel on proper operation and cleaning of production systems, including feeding procedures and proper use of equipment.

CERTIFICATION

❏ Sent a letter to your permitting authority, certifying that a BMP Plan was developed for your facility. Refer to Appendix F for an example of a certification letter.

* Be sure to define "routine" and "regular" (which can vary during the year) in your BMP Plan.

Appendix E2

Example BMP Plan

Example BMP Plan

3M & LJ Fish Farm
Prepared: September 30, 2004
NPDES Number: ID 1234567
Facility Manager: Bob Smith, 555-987-6543

[Note: this is an example BMP Plan and is not based on an actual facility.]

A. Description of Facility

3M & LJ's Fish Farm produces approximately 250,000 pounds of rainbow trout annually. The facility was originally constructed in 1976. It expanded in 1997 to include an off-line settling pond system. The facility currently has 12 100-foot long raceways, a small hatchery building, an office/shop, and an OLS pond for waste treatment (see Figure 1). The fish farm is located near Boise, Idaho. The facility has a non-consumptive water right for 14 cfs of water from Upper Springs. The facility has two discharge points, both of which go into Upper Creek.

B. Water Source

3M & LJ Fish Farm uses water from Upper Spring, which is a pure spring source with TSS levels generally measured at less than 2.0 mg/L (see historic DMRs). Aquatic vegetation grows around the spring head and the ditch leading to the raceways. An inflow trash rack screen at the facility is used to catch vegetation from the springs and ditch prior to entering the facility. The trash rack screen is cleaned at least daily to prevent vegetation from affecting the water flow to the facility. The spring and head ditch is manually cleaned twice a year to prevent build up of aquatic vegetation. The ditch has an adjustable head gate that controls the water flow to the facility from the spring area. The spring provides a constant supply of water to the facility and the water temperature remains nearly constant at 65 °F (± 20 °C).

C. Treatment System(s) Used

3M & LJ Fish Farm uses quiescent zones to capture solids in the production raceways. Solids are periodically removed and sent to an offline settling (OLS) pond for dewatering. Supernatant from the OLS pond discharges to Upper Creek.

At the downstream end of each raceway is a 20-foot long quiescent zone. The quiescent zone distance meets the minimum design criteria set forth in the *Idaho Waste Management Guidelines for Aquaculture Operations* for quiescent zone length. Each quiescent zone has a wastewater drain line connection that allows each to be vacuumed individually. The vacuum hose is attached to a slotted pipe that is 2 ft. long that serves a vacuum head. Floats are attached to the vacuum hose to prevent the hose from stirring up solids during cleaning events. Gravity transports the wastewater from the quiescent zone to the OLS pond for treatment and storage of settled solids. The delivery rate of wastewater to the OLS pond from the raceway or quiescent zone cleaning is 200 gpm.

The hatchery building is where trout eggs are hatched and the fish are raised up to a size where they can be moved outdoors to the production raceways to finish growing to market size. The troughs and small raceways in the hatchery building all have screened quiescent zones at their downstream ends.

The troughs, small raceways, and their corresponding quiescent zones are cleaned daily. The troughs and raceways all have a separate drain line that allows the cleaning wastewater to be diverted to the OLS pond. Water flow has been measured for the trough and small raceway quiescent zone drains and is 30 gpm and 75 gpm, respectively. Quiescent zone cleaning flows are recorded and used in the calculations for the discharge from the OLS pond. Water used in the hatchery building is diverted from the influent ditch below the weir and is discharged to the head ditch above the first raceways (see Figure 1).

The OLS pond has a design flow of 300 gpm. The dimensions of the OLS pond are 30 ft by 30 ft (surface area of 900 sq ft). The pond slopes to a maximum depth of 3.5 ft. Wastewater comes into the OLS pond from the gravity flow system pipe that spills onto the access ramp. This helps to distribute the flow across the width of the settling pond. Water leaves the pond through an 8 in. standpipe. The standpipe is attached to a 90° elbow that can swivel inside the pond. There is a collar around the standpipe that causes the water that is discharged to be pulled from 20 in. below the pond surface. The collar prevents floating materials from washing out of the pond. The water leaving the pond goes back to a box with a calibrated v-notched weir. The weir is used to verify flow rates through the OLS pond during cleaning events.

D. Other Information

3M & LJ Fish Farm uses an influent weir to measure flow for the facility. The weir is a calibrated suppressed rectangular weir and is located downstream from the trash rack screen to prevent debris from interfering with weir measurements. The weir is calibrated annually. The weir face and box area is swept clean prior to any measurements being taken. The staff gauge is placed along the weir box wall six times the head distance upstream of the weir crest. The weir has a 3/16 in. blade crest that falls off to a 45° angle to allow water to spring free of the blade. If the blade is nicked, bent, or rounded it is replaced. Weir calibration and testing curve validation are conducted annually. Immediately below the catch pool for the weir is the influent fish screen used to prevent fish from swimming out of the rearing areas and into the springs.

The raceways are grouped in four sections of three units and the groups are operated in series (see Figure 1). There is approximately 2.5 ft of drop between the first and second use raceways to allow for passive oxygen recharge of the raceway water. There is 3.5 ft of drop between the second and third raceway use. There is a 4.0 ft drop between the third and fourth raceway use. Between raceway sections the water falls onto a splashboard before entering the next lower section. The purpose of this splashboard is to break up the water stream leaving the upper raceway and expose as much surface area of the water to open air as possible to maximize the replenishment of dissolved oxygen levels in the raceway waters. After the fourth and final use, water falls 3.5 ft to a concrete pad before flowing into the tail ditch and off of the facility into Upper Creek, which accomplishes the same goal as the splashboards between raceway sets (i.e., it maximizes DO levels for wastewater entering Upper Creek).

E. Solids Control

1. Efficient feed management (to limit feed input to the minimum amount reasonably necessary to achieve production goals and sustain targeted rates of aquatic animal growth).

3M & LJ Fish Farm recognizes that fish feed management is critical in operating an environmentally friendly and profitable fish farm. Approximately 250,000 lb of trout are produced per year on about 300,000 lb of feed, at a conversion rate of 1.2. Feed used is produced from Best Feed for Fish and generally is composed of 42% protein, 16% fat, < 8% ash and less than 1.3% phosphorus. Feed contents change based on availability of constituents to the feed manufacturer. 3M & LJ Fish Farm keeps records of each shipment of feed received from the manufacturer, including the quantity and proximate analysis.

3M & LJ Fish Farm uses commercially available sinking extruded diets to feed our fish. Using extruded diets leads to the best feed conversion ratios, which minimizes the amount of waste generated by the facility. Specific quantities of feed are fed through demand feeders on each outdoor raceway depending on the quantity, size, and condition of the fish in that raceway. There are two demand feeders on each raceway. Demand feeders allow the fish to decide how much food they need and when they want to feed. This maximizes feeding opportunity and lowers feed conversions by providing a steady, stress free, feeding environment with little waste. Demand feeders are filled, at most, daily or as necessary. Prior to each filling, the feeders are also inspected for proper operation. Fish in the hatchery building are fed by hand several times per day.

Employees observe the feeding behavior of the fish throughout each shift. Fish that are not feeding well have their feed restricted until they are again feeding normally to prevent feed from being wasted and discharged.

When feeding the fish, our facility records the amount fed and estimates the number and weight of the trout being fed. From this information, we have been able to calculate FCRs to better manage feeding at our facility. Records of this information are kept in our office and are available upon request by contacting the facility manager. An example of the form we use is attached.

2. Procedures for routine cleaning of rearing units and offline settling basins.

Quiescent zones are vacuumed every two weeks and prior to fish grading or harvesting events. The screens in front of the quiescent zones are cleaned daily to remove moss (algae) and dead fish. Screens are cleaned to facilitate settling of biosolids from the raceway and to prevent blowouts, which occur when the screen is clogged and breaks from the water pressure. Fish that get into the quiescent zones are removed promptly when discovered. The troughs, small raceways, and their corresponding quiescent zones in the hatchery are cleaned daily.

Raceways above the quiescent zones are vacuumed before scheduled fish inventorying, grading or harvesting events.

The raceways are screened to prevent avian predators from eating the fish. The netting reduces indirect mortality from predators by reducing the incidence of disease at the facility. Healthy fish consume feed better, which prevents uneaten feed from going to waste, and are more active in the raceway, which allows accumulated biosolids to move more readily down the raceway to the quiescent zones, facilitating cleaning and faster removal of biosolids.

The OLS pond is harvested twice annually, in the spring and fall. When the OLS pond is harvested, the water in the pond is slowly decanted by removing the collar from around the standpipe and slowing rotating the standpipe on the 90° elbow to gradually lower the water level in the pond. Once the pond is decanted, a tractor is driven into the pond and the slurry is stirred to a uniform consistency to allow for pumping. The sludge is pumped from the OLS pond into a "honey wagon," which takes it to a field for land application. Solids content of the slurry varies between 6 % and 12 %.

Sludge and slurry that have been collected in the OLS pond are recycled by land application to nearby cropped fields. Farmers that accept the slurry agree to disc it under within 24 hours of application and prior to any irrigation water being applied to the field. All land application is done in such a manner as to prevent the materials from entering surface or groundwaters. The dates, locations, and amounts of slurry that are taken offsite for land application are kept in a record.

Examples of the forms we use to track cleaning and land application of waste are attached. Records of this information are kept in our office and are available upon request by contacting the facility manager.

3. Procedures for inventorying, grading, and harvesting aquatic animals (that minimize discharge of accumulated solids).

Raceways above the quiescent zones are vacuumed before any scheduled inventorying, grading, or harvesting events to prevent unnecessary disturbance and subsequent discharge of biosolids from the raceways.

4. Remove and dispose of aquatic animal mortalities properly on a regular basis to prevent discharge to waters of the United States (except where authorized by your permitting authority in order to benefit the aquatic environment).

Raceways at our facility are screened to prevent avian predators from eating the fish. This benefits the waste management on the farm by reducing direct mortality to injured fish. The netting reduces indirect mortality by reducing the incidence of disease at the facility. When fish mortality does occur, carcasses are promptly removed (at a minimum – daily). Fish carcasses are typically composted on site. Mortalities from the hatchery are also disposed of with the raceway mortalities. In the event of a significant problem that leads to a large number of mortalities, a local rendering company is called to haul the mortalities away.

Mortalities generally range from 1% to 7% of fish on hand, now that the raceways are screened, and depending on the disease and timing of the disease outbreak.

Our facility tracks carcass removal to improve facility management. When we encounter fish mortalities, we record the number of fish that died, their approximate weight, the group/age, and how we disposed of them. Records of this information are kept in our office and are available upon request by contacting the facility manager. An example of the form we use is attached.

F. Material Storage

1. Proper storage of drugs, pesticides, and feed to prevent spills that may result in the discharge to waters of the United States.

To ensure proper storage of drugs, pesticides, and feed at 3M & LJ Fish Farm, employees have been instructed on the importance of proper handling of these substances through the facility-training program.

Bagged feeds are stored in the shop area and are used on a "first in, first out" basis to prevent lengthy storage of feed. Use of fresh diets improves dietary efficiency. No feed is used if it has exceeded the storage period recommended by the manufacturer. This rarely occurs, but when outdated feed is found, it has to be taken to the local landfill for disposal. The largest diets are purchased in bulk and stored in feed bins. Feed can be poured from the bins and fines screened off before the feed is put in the demand feeders. All fines are collected and sent back to the manufacturer for repelleting. Since converting to extruded pellets, the volume of fines is typically less than 50 lb per month. All the demand feeders are set up with a windshield to prevent the undesired release of feed on windy days.

All drugs, disinfectants, chemicals, and pesticides are stored in a cabinet in the office building in their original containers. The chemical cabinet is in a dry well-ventilated place, away from water, and with no floor drains. Employees have been instructed to keep container lids secure at all times. All liquid materials at our facility are stored in a containment system to prevent possible spills from running off. Every month, we inspect and maintain the storage areas and equipment to prevent spills.

2. Procedures for properly containing, cleaning, and disposing of any spilled materials.

3M & LJ Fish Farm has developed a spill response and prevention plan to ensure that our facility properly contains, cleans, and disposes of spilled materials. The plan provides details on measures to stop the source of a spill, contain the spill, clean up the spill, dispose of contaminated materials, and train personnel to prevent and control future spills. The plan is reviewed during our required annual employee-training program.

Our spill response and prevention plan is a procedural handbook that identifies individuals responsible for implementing the plan; defines safety measures to be taken with each kind of waste; emphasizes that spills must be cleaned up promptly; specifies how to notify appropriate authorities, such as the fire department for assistance; states procedures for containing, diverting, isolating, and cleaning up the spill; and describes spill response equipment to be used, including safety and cleanup equipment.

Our plan encourages the use of shop rags (for small spills) and absorbent snakes (for large spills), rather than water. For non-hazardous materials, we would send the rags out for cleaning and throw away the absorbent snakes. For hazardous materials, we would dispose of the cleanup materials according to our state's guidelines. A copy of the plan is kept in our office and is available upon request by contacting the facility manager.

We also keep track of information about any spills at our facility to try to prevent any future spills. This information is kept with other records for the facility in the office. An example of the form we use to track spills is attached.

☐. Mai☐te☐a☐☐e

1. Routinely inspect production systems and wastewater treatment systems to identify and promptly repair damage.

We inspect the raceways and quiescent zones every day. More specifically, we check that all of the drain structure parts are functioning properly; that valves and other critical drain components are working properly; and that there are no broken parts. If any parts are broken, we repair them immediately. We check the raceways to make sure they are structurally sound; repair cracks as necessary; and check that all plumbing components are installed and working properly. Finally, we check the quiescent zones for proper function; inspect drains for clogging; and make sure that all settling basins are working properly.

We also check equipment used at the facility every day. We routinely inspect oxygen equipment, filters, and heaters that maintain optimal growing conditions in the hatchery. We also test the demand automatic feeders periodically (weekly) to ensure they are delivering the proper amounts of feed; check demand feeders for proper operation and adjust as necessary; and inspect all feed storage areas to make sure the feed is free from rodents and insects and that no excess moisture or water leaks are present to prevent mold.

We keep track of the inspections of production and wastewater treatment systems at our facility in a log. Records of this information are kept in our office and are available upon request by contacting the facility manager. An example of the log is attached.

2. Regularly conduct maintenance of production systems and wastewater treatment systems to ensure their proper function.

We perform maintenance of parts and equipment when our routine inspections determine that a repair is necessary. We also perform maintenance on equipment that requires periodic maintenance or adjustment, based on the manufacturer's recommendations. For example, demand feeders need to be constantly adjusted to the conditions of the facility to maximize feeding efficiency. Employees immediately correct any feeders discovered to be out of adjustment (feeding too freely or jammed up). We record maintenance in the same log where we record inspections of production and wastewater treatment systems. An example of this form is attached.

☐. ☐e☐or☐☐☐ee☐i☐g

1. Maintain records for aquatic animal rearing units documenting feed amounts and estimates of the numbers and weights of aquatic animals in order to calculate representative feed conversion ratios.

We keep records in the office of the amount of feed we feed our fish daily. We also record estimates of the number and weight of trout at our facility at the same time we feed the fish. From this information, we have been able to calculate FCRs to better manage feeding at our facility. All records are maintained and updated as information is collected. For example, feeding records are updated daily. All records are available upon request by contacting the facility manager. Examples of the forms we use are attached.

2. Keep records documenting frequency of cleaning, inspections, maintenance, and repairs.

We keep records in the office of the frequency of cleaning, inspections, maintenance, and repairs at our facility. All records are maintained and updated as information is collected. All records are available upon request by contacting the facility manager. Examples of the forms we use are attached.

☐ ☐rai☐i☐g

1. Train all relevant personnel in spill prevention and how to respond in the event of a spill to ensure proper clean-up and disposal of spilled materials.

All new employees are required to attend training on spill prevention and response. Current employees must attend refresher training once every year. Our facility has developed a spill prevention and response plan, which is covered at employee training. The plan was described in more detail in Section F above. The plan is kept in our office, is accessible to all employees, and is available upon request by contacting the facility manager. We also keep Material Safety data sheets for all chemicals used at the facility within a binder in the chemical cabinet. Employees have been trained on how to use these sheets.

2. Train personnel on proper operation and cleaning of production and wastewater treatment systems, including feeding procedures and proper use of equipment.

All new employees are required to attend training on proper operation and cleaning of production systems and wastewater treatment systems at our facility. Management reviews employee performance in operating the facility and provides additional training for those not operating the facility according to the Standard Operating Procedures.

☐. ☐iagra☐ or Ma☐

A diagram/map of the facility (to illustrate the layout of the operation) is included at the end of the BMP plan as Figure 1.

☐. ☐e☐ie☐ a☐☐ ☐☐☐or☐e☐e☐t o☐t☐e ☐M☐ ☐la☐

We, the facility manager and the individuals responsible for implementing the BMP plan, have reviewed and endorsed this BMP plan.

_____	_____
(Facility Name)	(NPDES #)
_____	_____
(Facility Manager – Printed Name)	(Facility Manager – Signature)
_____	_____
(Other Individual – Printed Name & Title)	(Other Individual – Signature)

☐. ☐erti☐☐i☐g t☐e ☐M☐ ☐la☐ ☐it☐ t☐e ☐er☐itti☐g ☐☐t☐orit☐

A copy of the BMP plan is kept in our office. It is available to any employee at our facility, EPA, and any state environmental agency upon request by contacting the facility manager.

A signed letter has been sent to our permitting authority stating that our facility has developed a BMP plan.

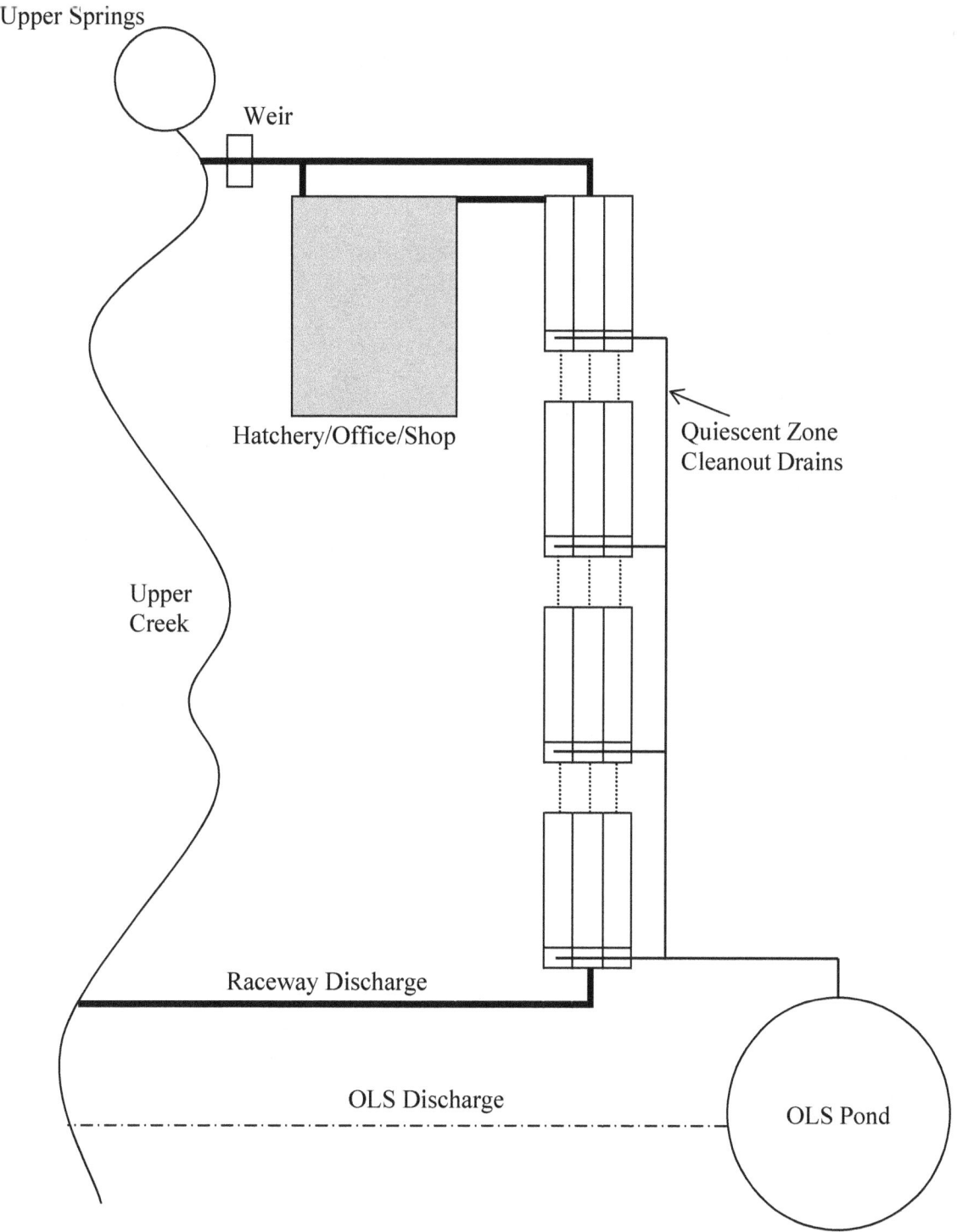

Figure 1: Facility Diagram of 3M & LJ Fish Farm

Feed Conversion Ratios Log

Facility Name: 3M & LJ Fish Farm   **NPDES Permit Number:** ID 1234567

Instructions: Fill in this form with feeding information so as to keep track of feeding and to calculate/track feed conversion ratios. FCRs are calculated with the following equation:

$$\frac{\text{Dry weight of feed applied}}{\text{Wet weight of fish gained}}$$

Date (start date end date)	Description of Group	Total Feed Amounts (Estimate)	Weights of Animals (start weight end weight)	Weight Gained	Calculated FCR
3/1/04 10/1/04	Rainbow trout stockers	20,775 lbs	100 lbs 17,528 lbs	17,428 lbs	1.19

Cleaning Log

Facility Name: 3M & LJ Fish Farm   **NPDES Permit Number:** ID 1234567

Instructions: Record all cleaning performed on production and/or wastewater system components.

Date Cleaned	Cleaner Initials	Description of Component Cleaned	Notes About Cleaning
10/1/04	ML	QZ raceways 1-10	Cleaned QZs and dam boards; checked and cleaned screens, removing moss and dead fish.

OLS Waste Land Application Log

Facility Name: 3M & LJ Fish Farm   **NPDES Permit Number:** ID 1234567

Instructions: Record information about land application of OLS waste in the form below. This is an example of a state requirement that is not required by the ELGs.

Date	Initials	Location	Nutrient Analysis	Volume Applied	Notes
10/1/04	ML	Crop Field 1	2.5%% nitrogen 3.0% phosphorus 0.1% potassium 5.0% calcium 1.0% magnesium	10,000 gallons	No problems encountered.

CARCASS REMOVAL LOG

Facility Name: 3M & LJ Fish Farm   **NPDES Permit Number:** ID 1234567

Instructions: Record all mortalities observed on a daily basis.

Date	Initials	System/Group of Animals	# of Mortalities	Approx. Weight	Disposal Method	Notes
10/1/04	ML	Rainbow trout – R1–12	20	11 lbs	Composted	All from R4; appear to be from an infection – closely monitoring remaining fish

EXAMPLE SPILLS AND LEAKS LOG

Facility Name: 3M & LJ Fish Farm   **NPDES Permit Number:** ID 1234567

Instructions: Fill in this form with information about any spills or leaks.

Date (mm/dd/yy)	Spill or Leak	Location (as indicated on a site map)	Type of Material & Quantity	Source (if known)	Reason	Amount of Material Recovered	List of Preventative Measures Taken	Initials
10/4/04	Spill	Hatchery floor	Formalin	Storage drum	Top was not secured and the drum was knocked over	20 gallons	Spoke to all employees about the importance of securing lids to storage containers	ML

Instructions: Use this page to enter any important notes about the spills from the previous sheets.

Date of spill or leak (mm/dd/yy)	Notes
10/4/04	All employees have been instructed on the importance of securing container lids. Employees were also instructed to double-check that container lids are secured before proceeding with applications.

EPA-821-B-05-001

STRUCTURAL MAINTENANCE
INSPECTION AND MAINTENANCE LOG

Facility Name: 3M & LJ Fish Farm   **NPDES Permit Number:** ID 1234567

Production or Wastewater Treatment System: Raceway 1

Instructions: Fill in all routine inspections and regular maintenance of production systems and wastewater treatment systems in the table below. Use a separate form for each production system and/or wastewater treatment system.

Date Inspected	Inspector Initials	Notes (Note any problems found and maintenance performed)	Date Maintenance Performed
10/1/04	ML	The screens at the end of raceways 5 and 7 were loose; secured the screen.	10/2/04

Appendix F

BMP Certification Form

BMP Certification Form

Facility Name:_____   NPDES Permit Number:_____

Printed Name:_____

Title (owner, operator, etc.):_____

Date the BMP Plan was developed:_____

I certify that a BMP plan was developed for:_____

(name of facility)

A copy of the BMP plan is available for inspection at the following address:

_____

_____

_____

Signature:_____   Date:_____

* Note: This is only an example of what a certification form could look like.

Appendix G

State BMP Programs

State BMP Programs

A number of states, including Alabama, Arizona, Arkansas, Florida, Hawaii, and Idaho, were found to have recommended BMPs for AAP. In addition, BMPs have also been developed for specific types of aquatic species. BMPs are addressed in manuals or regulations, depending on the state. Data were collected from in-house resources and through Internet research. An example of technical guidance on BMP development is *Best Management Practices for Flow-through, Net Pen, Recirculating, and Pond Aquaculture Systems* (Tucker et al., 2003). This guidance document provides examples of existing BMP plans and state regulations, as well as technical information that can be used in facilities' BMP plan development. Information is provided for four production system types and ranges from guidance on site selection, to solids and feed management, to facility operation and maintenance.

Alabama

Dr. Claude Boyd and his colleagues, with funding from the Alabama Catfish Producers (a division of the Alabama Farmers Federation), have developed a set of BMPs for aquaculture facilities in Alabama. The BMPs are described in a series of guide sheets that have been adopted by USDA's Natural Resources Conservation Service (NRCS) to supplement the Service's technical standards and guidelines (Auburn University and USDA, 2002). The NRCS technical standards are intended to be referenced in Alabama Department of Environmental Management rules or requirements that are promulgated for aquaculture in Alabama. The guide sheets address a variety of topics, including

reducing storm runoff into ponds, managing ponds to reduce effluent volume, erosion control in watersheds and on pond embankments, settling basins and wetlands, and feed management.

Arizona

Arizona Aquaculture BMPs addresses treatment and discharge of aquaculture effluents containing nitrogenous wastes and closing of aquaculture facilities when they cease operation (Fitzsimmons, 1999).

Use of these BMPs is intended to minimize the discharge of nitrates from facilities without being too restrictive for farm operations. The draft document *Arizona Aquaculture BMPs* describes BMPs that can minimize nitrogen impacts from aquaculture facilities. A list of information resources is also provided for additional information about Arizona aquaculture and BMPs (Fitzsimmons, 1999). The BMPs are available at: http://ag.arizona.edu/azaqua/bmps.html.

Arkansas

The Arkansas Bait and Ornamentals Fish Growers Association (ABOFGA, n.d.) developed a list of BMPs to help its members make their farms more environmentally friendly. More specifically, the Association provides a set of BMPs that help to conserve water, reduce effluent, capture solids, and manage nutrients. Members may voluntarily agree to adopt the BMPs on their farms (ABOFGA, n.d.).

Florida

Florida's aquaculture certificate of registration and BMP program requires any person engaging in aquaculture to be annually certified by the Florida Department of Agriculture and Consumer Services and to follow BMPs established by the Department (Chapter 597, Florida Aquaculture Policy Act, Florida Statutes). *Aquaculture Best Management Practices*, a manual prepared by the department, establishes BMPs for aquaculture facilities in Florida. By legislative mandate, the BMPs in the manual are intended to preserve environmental integrity, while eliminating cumbersome, duplicative, and confusing environmental permitting and licensing requirements. When these BMPs are followed, aquaculturists meet the minimum standards necessary for protecting and maintaining offsite water quality and wildlife habitat (FDACS, 2000). Additional information is available from Florida's Division of Aquaculture website at http://www.FloridaAquaculture.com.

Georgia

Agriculture enterprises such as fish farms are required to conduct activities consistent with Best Management Practices (BMP's) established by the Georgia Department of Agriculture. BMP's are management strategies for control and abatement of nonpoint source pollution resulting from agriculture. The manual *Agricultural Best Management Practices for Protecting Water Quality in Georgia* provides information on using and maintaining BMP's. The manual is available from the Georgia Soil and Water Conservation Commission (Georgia Department of Natural Resources, 2003).

Hawaii

Hawaii developed a practical BMP manual to assist aquaculture farmers in managing their facilities more efficiently and complying with discharge regulations. The manual, *Best Management Practices for Hawaiian Aquaculture* (Howerton, 2001), is available from the Center for Tropical and Subtropical Aquaculture. A copy of the manual is available at: http://www.ctsa.org/upload/publication/CTSA_148631672853284080260.pdf.

Idaho

In combination with site-specific information, *Idaho Waste Management Guidelines for Aquaculture Operations* can be used to develop a waste management plan to meet water quality goals. Such a waste management plan would address Idaho's water quality concerns associated with aquaculture in response to the Clean Water Act and Idaho's Water Quality Standards and Wastewater Treatment Requirements. The manual is also intended to assist aquaculture facility operators in developing BMPs to maintain discharge levels that do not violate the state's water quality standards (IDEQ, n.d.). The manual is available for download at: http://www.deq.state.id.us/water/prog_issues/agriculture/aquaculture.cfm.

Louisiana

The LSU AgCenter has published a guidance manual, *Aquaculture Production Best Management Practices (BMPs)*, which provides a list of BMPs that can help producers to conserve soil and protect water and air resources by reducing pollutants. The manual is available at http://www.lsuagcenter.com/en/environment/conservation/bmps/aquaculture+production+best+management+practices.htm.

Missouri

Missouri has published the following guidance document to answer typical questions that aquaculture facility owners or operators may have: *Missouri Aquaculture Environmental and Regulatory Guide: A Guide to Regulatory Compliance, Sources of Information and Assistance and Answers to Environmental Questions for Aquaculture Businesses in Missouri*. The guide provides basic information about regulatory requirements and suggestions for protecting operators and owners of aquaculture facilities, their workers and the environment through pollution prevention. Each guide sheet in the publication deals with a separate issue, such as pollution prevention, dead fish disposal, backflow prevention, drug use, and preventing fish diseases. A copy of the guide is available at: http://www.dnr.mo.gov/pubs/pub513.PDF.

Ohio

The *Ohio Pond Management Handbook*, created by the Department of Natural Resources' Division of Wildlife, provides guidance to pond or small lake owners on pond management issues. Owners of new ponds, owners of old ponds, or landowners who plan to build a pond are given guidance on how to best manage fish stocks, aquatic vegetation, fish health, and surrounding pond wildlife (Ohio DNR, 1996). The guide is available for download at: http://www.dnr.state.oh.us/wildlife/PDF/pon dmgt.pdf.

West Virginia

West Virginia University, Extension Service has developed a guide, as part of their Aquaculture Information Series, for managing aquaculture waste entitled *Waste Management in Aquaculture*. The guide discusses BMPs, and their associated costs,

that may be used to reduce aquaculture waste. A copy of the guide is available at: http://www.wvu.edu/~agexten/aquaculture/ waste02.pdf. Other aquaculture-related fact sheets and documents are available at: http://www.wvu.edu/~agexten/aquaculture/f actsht.htm.

Wisconsin

As of July 2005, the University of Wisconsin is finishing completing a draft version of a BMP user manual for aquaculture for Wisconsin and the Great Lakes Region. The draft document will be reviewed by various agencies before it is made available to the public.

Other Aquaculture Documents

BMPs have also been developed for specific species, including shrimp, hybrid striped bass, and trout. The Global Aquaculture Alliance, in *Codes of Practice for Responsible Shrimp Farming*, has compiled nine recommended codes of practice that are intended to serve as guidelines for parties who want to develop more specific national or regional codes of practice or formulate systems of BMPs for use on shrimp farms. These codes of practice address a variety of topics, including mangroves, site evaluation, design and construction, feeds and feed use, shrimp health management, therapeutic agents and other chemicals, general pond operations, effluents and solid wastes, and community and employee relations (Boyd, 1999). The purpose of the document is to provide a framework for environmentally and socially responsible shrimp farming that is voluntary, proactive, and standardized. The document also provides a background narrative that reviews the general processes involved in shrimp farming and the environmental and social issues facing the industry (Boyd, 1999).

The Hybrid Striped Bass Industry: From Fish Farm to Consumer is a brochure that provides guidance to new and seasoned farmers in the proper handling of fish from the farm to the consumer. Although the brochure is primarily geared toward providing quality fish products to consumers, the information it provides about the use of drugs and chemicals, including pesticides and animal drugs and vaccines, could be used to benefit the environment (Jahncke et al., 1996).

The Trout Producer Quality Assurance Program of the U.S. Trout Farmer's Association (USTFA) is a two-part program that emphasizes production practices that enable facilities to decrease production costs, improve management practices, and avoid any possibilities of harmful drug or other chemical residues in fish. Part 1 discusses the principles of quality assurance, and Part 2 provides information about the highest level of quality assurance endorsed by the USTFA. Although the program addresses a variety of subjects related to trout production, the discussion on waste management and drugs and chemicals can be applied to protecting the environment (USTFA, 1994).

References

ABOFGA. No Date. *Best Management Practices (BMP's) for Baitfish and Ornamental Fish Farms*. Arkansas Bait and Ornamental Fish Growers Association, in cooperation with the University of Arkansas at Pine Bluff, Aquaculture/Fisheries Center.

Boyd, C.E. 1999. *Codes of Practice for Responsible Shrimp Farming*. Global Aquaculture Alliance, St. Louis, Missouri. 48 pages. Copies of this publication are available by contacting the Global Aquaculture Alliance at P.O. Box 510799,

St. Louis, Missouri, 63151-0799, 314-416-9500, Fax: 314-416-9500, E-mail: homeoffice@gaalliance.org, Web page: http://www.GAAlliance.org.

DEQ. No Date. *Idaho Waste Management Guidelines for Aquaculture Operations*. Idaho Division of Environmental Quality. Available for download at <http://www.deq.state.id.us/water/prog_issu es/agriculture/aquaculture.cfm>. Accessed September 2004.

FDACS. 2000. *Aquaculture Best Management Practices*. Florida Department of Agriculture and Consumer Services, Division of Aquaculture, Tallahassee, Florida.

Fitzsimmons, K. 1999. *Draft: Arizona Aquaculture BMPs*. Arizona Department of Environmental Quality. Available for download at <http://ag.arizona.edu/azaqua/bmps.html>. Accessed September 2004.

Georgia Department of Natural Resources. 2003. *A Summary of Georgia's Licenses and Permits for Aquaculture*. <http://georgiawildlife.dnr.state.ga.us/assets/ documents/Aqua_Sum_Oct03.pdf>. Accessed October 2004.

Howerton, R. 2001. *Best Management Practices for Hawaiian Aquaculture*. Center for Tropical and Subtropical Aquaculture, University of Hawaii Sea Grant Extension Services, Publication No. 148. <http://www.ctsa.org/upload/publication/CT SA_148631672853284080260.pdf>. Accessed September 2004.

Jahncke, M.L, T.I.J. Smith, and B.P, Sheehan. 1996. *The Hybrid Striped Bass Industry: From Fish Farm to Consumer*. Preparation of this brochure was supported by the U.S. Department of Agriculture, FISMA Grant No. 12-25-G-0131, U.S. Department of Commerce, National Marine Fisheries Service, South Carolina Department of Natural Resources, and South Carolina Department of Agriculture.

Ohio DNR. 1996. *Ohio Pond Management Handbook.* Ohio Department of Natural Resources, Division of Wildlife. <http://www.dnr.state.oh.us/wildlife/PDF/pondmgt.pdf>. Accessed June 2005.

USTFA. 1994. *Trout Producer Quality Assurance Program.* U.S. Trout Farmer's Association.

Appendix H

National Association of State Aquaculture
Coordinators (NASAC), Cooperative Extension
Services, and Sea Grant Information

National Association of State Aquaculture Coordinators (NASAC), Cooperative Extension Services, and Sea Grant Information

Since this guidance document was completed, contacts for NASAC, Cooperative Extension Services and Sea Grant Programs may have been updated. If any of the links do not work, please visit the following websites for updated contacts:

□A□A□

http://www.marylandseafood.org/aquaculture/nasac.php

□oo□□ra□i□□ □□□□nsion

http://www.csrees.usda.gov/qlinks/partners/state_partners.html

□□a □ran□

http://www.nsgo.seagrant.org

A□A□A□A

NASAC

John Gamble
Marketing & Economics Division
Department of Agriculture and Industries
PO Box 3336
Montgomery AL 36109-0336
(334)240-7245
(334)240-7270 FAX
ageconadm@agi.state.al.us

Jimmy Carlisle
Catfish, Poultry and International Trade
Divisions
Alabama Farmers Federation
PO Box 11000
Montgomery AL 36191-0001
(334)613-4214
(334)284-3957 FAX
jcarlisle@alfafarmers.org

Cooperative Extension

The Department of Fisheries and Allied Aquaculture is responsible for extension activities in aquaculture production and marketing, managing fish populations in large and small impoundments, aquatic ecology, and recreational fisheries. Department experts work with county extension agents throughout the state to conduct educational programs and provide up-to-date, practical information to interested clients. More information about the Department is available at:

http://www.ag.auburn.edu/dept/faa/extension

Sea Grant

Mississippi-Alabama Sea Grant Consortium (MASGC) is an organization of nine universities and laboratories dedicated to activities that foster the conservation and sustainable development of coastal and marine resources in Mississippi and Alabama. Additional information about MASGC is available at:

http://www.masgc.org

ALASKA

NASAC

Guyla McGrady
Department of Natural Resources
Division of Land
550 W 7th Ave, Suite 900C
Anchorage AK 99501-3577
907-269-8543
907-269-8913 FAX
guyla_mcgrady@dnr.state.ak.us

Cynthia Pring-Ham
Alaska Department of Fish and Game
P O Box 25526
Juneau AK 99802-5526
(907)465-6150
(907)465-4168 FAX
cynthia_pring-ham@fishgame.state.ak.us

Cooperative Extension

The University of Alaska Fairbanks Cooperative Extension has had aquaculture-related initiatives through its 4-H programs. Information about University of Alaska-Fairbanks Cooperative Extension efforts can be found at:

http://www.uaf.edu/coop-ext

Sea Grant

The Marine Advisory Program (MAP) works to promote positive development, use and conservation of Alaska's aquatic and marine resources. The Aquaculture section of the MAP attempts to promote the aquaculture industry in Alaska through improvement of species diversity, production, quality, management, marketing opportunities, and public awareness of the industry. More information about the Marine Advisory Program can be obtained at:

http://www.uaf.edu/map/about.html

AMERICAN SAMOA

NASAC

Ta'alo P. Lauofo
Department of Agriculture
Pago Pago, American Samoa 96799

Cooperative Extension

American Samoa Community College's Cooperative Extension Service offers community-based educational programs and projects to enhance individual and group decision-making towards improved living. Extension works closely with farmers, homemakers and youth, as well as government and civic agencies. The extension agents use discoveries made by the research division to improve the quality of life for individuals and the community.

Extension programs are offered in agriculture, consumer family sciences, 4-H youth, and forestry. No specific information concerning aquaculture for the state's cooperative extension service was found. Specific information about the cooperative extension service is available at:

http://www.ascc.as/academicssupportcnrp.htm

Sea Grant

American Samoa Community College does have a Sea Grant Program. However, no information or website was available.

ARIZONA

NASAC

Richard Willer
Department of Agriculture
1688 West Adams
Phoenix AZ 85007-2617
(602)542-4293
(602)542-4290 FAX
rwiller@azda.gov

Cooperative Extension

The University of Arizona Cooperative Extension Service acts as a conduit for information on aquaculture issues. Information available at this website includes links to publications, educational activities including the high school aquaculture program, and other resources within the State. Additional information about University of Arizona Aquaculture is available at:

http://ag.arizona.edu/azaqua

Sea Grant

There is no Sea Grant Program for the State of Arizona.

ARKANSAS

NASAC

Ted McNulty
Arkansas Development and Finance Authority
423 Main Street Suite 500
Little Rock AR 72203
(501)682-5849
(501)682-5893 FAX
tmcnulty@adfa.state.ar.us

Cooperative Extension

The Aquaculture/Fisheries Center at the University of Arkansas Pine Bluff hosts information on Extension activities. The Center provides educational materials such as newsletters, videos, demonstrations and training. Some specialized programs have included publication of the quarterly newsletter "Aquaculture Aquafarming," and development of Spanish-English training curriculum and training. The goal of the Center is to provide relevant, timely information through research and extension programs. Further information about the Aquaculture/ Fisheries Center can be obtained at:

http://www.uaex.edu/aqfi

Sea Grant

Arkansas does not have a Sea Grant Program.

CALIFORNIA

NASAC

Bob Hulbrock
Department of Fish and Game
1812 Ninth Street
Sacramento CA 95814
916-445-4034
916-445-4044 FAX
rhulbrock@dfg.ca.gov

Cooperative Extension

The University of California, Davis Animal Science Extension is linked to about 30 county-based Cooperative Extension livestock and dairy farm advisors. The campus-based specialists are responsible for the program areas of waste management, livestock systems management, aquaculture, and dairy management and health. The Animal Science Extension provides Aquaculture Document Database, an online source for documents related to Aquaculture both in California and throughout the United States. Information about the Animal Science Extension is available at:

http://animalscience.ucdavis.edu/extension/Aquaculture.htm

Sea Grant

The California Sea Grant College program focuses its work on aquaculture issues of research and development. Abstracts of research documents pertaining to this subject are available online. The Southern California Sea Grant has identified aquaculture as a research need and places development of aquaculture as a topic under the "Urban Coasts" theme. More information about the California Sea Grant College program's aquaculture research is available at:

http://www-csgc.ucsd.edu/RESEARCH/SgResearchIndx.html

Information about the Southern California Sea Grant can be found at:

http://www-csgc.ucsd.edu

COLORADO

NASAC

Jim Rubingh
Department of Agriculture
700 Kipling Suite 4000
Lakewood CO 80215-8000
(303)239-4117
(303)239-4125 FAX
jim.rubingh@ag.state.co.us

Cooperative Extension

The Colorado State University Extension has one aquaculture expert within the faculty of the College of Natural Resources, who has expertise about both sport fishing pond management and fish farming. More information about the Colorado State University Extension can be found at:

http://www.ext.colostate.edu

Sea Grant

Colorado does not have a Sea Grant Program.

CONNECTICUT

NASAC

David Carey
Bureau of Aquaculture & Laboratory
Connecticut Department of Agriculture
PO Box 97
Milford CT 06460
(203)874-0696
(203)783-9976 FAX
DEPT.AGRIC@SNET.NET

Cooperative Extension

The University of Connecticut Cooperative Extension seeks to strengthen profitability and to increase agriculture as a major sector in the State economy. Work is done in a variety of areas, including aquaculture, to update production and improve management skills. Additional information about the

University of Connecticut Cooperative Extension is located at:

http://www.canr.uconn.edu/ces

Sea Grant

Connecticut Sea Grant seeks to create possibilities within the State for new jobs through sustainable, environmentally friendly intensive aquaculture. The Sea Grant promotes research projects to further this end. Additional information about Connecticut Sea Grant College's aquaculture initiatives can be found at:

http://www.seagrant.uconn.edu/aqua.htm

DELAWARE

NASAC

Bruce Walton
Department of Agriculture
2320 South Dupont Highway
Dover DE 19901-5515
(302)698-4503
(302)697-4463 FAX
brucew@state.de.us

Cooperative Extension

The University of Delaware Cooperative Extension provides local communities with information on agricultural issues, financial tools, and other community educational materials. No information concerning aquaculture for the state's cooperative extension service was available for Delaware. Complete information about the University of Delaware Cooperative Extension can be found at:

http://ag.udel.edu/extension

Sea Grant

The University of Delaware Sea Grant Marine Advisory Service acts as a source of information-sharing among researchers, resource managers, business people and private citizens in an effort to foster the wise use, conservation, and development of marine resources. More information about the University of Delaware Sea Grant Marine Advisory Service's programs in aquaculture is available at:

http://www.ocean.udel.edu/mas/aquaculture/aquaculture.html

FLORIDA

NASAC

Sherman Wilhelm
Florida Department of Agriculture
Division of Aquaculture
1203 Governor's Square Blvd. Fifth Floor
Tallahassee FL 32301
(850)488-4033
(850) 410-0893 Fax
wilhels@doacs.state.fl.us
http://www.FloridaAquaculture.com

Cooperative Extension

The Institute of Food and Agricultural Sciences serves as the coordinating body for Extension Service activities and information. The Extension Service is a partnership of county, state, and federal government, which serves the citizens of Florida by providing information and training on a wide variety of topics. This site provides contact information for experts specializing in fisheries, aquaculture and pond management, and also has links to relevant publications. More information about extension activities of the Institute of Food and Agricultural Sciences is available at:

http://fishweb.ifas.ufl.edu/ExtensFac.htm

Sea Grant

Florida Sea Grant College efforts in aquaculture seek to promote sustainable production of aquatic species through education and research. Further details of Florida Sea Grant College efforts can be found at:

http://www.flseagrant.org

GEORGIA

NASAC

Ted Hendrickx
Wildlife Resources Division - Fisheries
2123 US Highway 278 SE
Social Circle GA 30025
770-918-6418
706-57-3040 FAX
ted_hendrickx@dnr.state.ga.us

Cooperative Extension

The Warnell School of Forest Resources serves as the conduit for information about forestry, fisheries, wildlife and conservation for the State of Georgia. One goal of the Warnell School is to provide information to the citizens of Georgia and the US so that they might be able to reach informed decisions about personal objectives and societal issues involving forestry and forest products, wildlife, aquaculture and fisheries and related natural resources. Aspects of these efforts include: education through technology transfer, translation and synthesis of research results and other information, discussion and explanation of public policy issues and science-based evaluations and recommendations. Further information about the Warnell School of Forest Resources can be found at:

http://www.forestry.uga.edu

Sea Grant

The University of Georgia Sea Grant College collaborates with the University of Georgia Marine Extension Service to provide information about current research in aquaculture. Additional information about the Georgia Sea Grant College is available at:

http://www.marsci.uga.edu/gaseagrant

GUAM

NASAC

Jeff Tellock
Guam Department of Commerce
Guam Aquaculture Development & Training
102 M Street
Tiyan GU 96913
(671)734-3011/7327
(671)477-9031 FAX

Cooperative Extension

The University of Guam's Cooperative Extension Service focuses on agriculture and natural resources. No specific information related to aquaculture from the state's Cooperative Extension Service was found. Additional information about the Service is available at:

http://www.uog.edu/cals/site/extension.html

Sea Grant

The mission of the University of Guam's Sea Grant Program is to optimize the sustainable use of the ocean's resources, to protect the delicate ecosystems that exist and to prevent any hazards and degradation of the natural resources through research, education, and advisory support; to increase our understanding of the balance of sustainability and protection of the environment that exist within the Western Pacific region; and to make significant

contributions for the benefit of Guam, the Western Pacific, and the Nation. No website was available for the University of Guam's Sea Grant Program.

HAWAII

NASAC

John Corbin
Aquaculture Development Program
Department of Land and Natural Resources
1177 Alakea Street Room 400
Honolulu HI 96813
(808)587-0030
(808)587-0033 FAX
info@hawaiiaquaculture.org

Cooperative Extension

The University of Hawai'i Center of Tropical Agriculture and Human Resources conducts research and provides the public with information on a variety of topics including community development, environmental issues, and commercial production. Aquaculture is listed as one area of focus, but further details of programs and research were not found on the website. More information about the University of Hawai'i Center of Tropical Agriculture and Human Resources is located at:

http://www.ctahr.hawaii.edu/ctahr2001/Extension/ExtMain.html

Sea Grant

Promotion of sustainable aquaculture is one research goal of the Sea Grant in Hawaii. More information about the University of Hawaii Sea Grant is available at:

http://www.soest.hawaii.edu/SEAGRANT

IDAHO

NASAC

Dan Crowell
Division of Animal Industries
Department of Agriculture
PO Box 7249
Boise ID 83707
(208)332-8540
(208)334-4062 FAX
dcrowell@agri.state.id.us

Cooperative Extension

The Aquaculture Research Institute provides students with educational and research opportunities relating to aquaculture and serves as an active educational outreach program within the state of Idaho. The Institute promotes, supports, and coordinates aquaculture research activities at the University of Idaho and throughout the state. Additional information about the Institute can be found at:

http://www.webs.uidaho.edu/aquaculture

Sea Grant

Idaho does not have a Sea Grant program.

ILLINOIS

NASAC

Ms. Delayne Holsapple Reeves
Department of Agriculture
State Fairgrounds
PO Box 19281
Springfield IL 62794-9281
(217)524-9129
(217)524-5960 FAX
dreeves@agr.state.il.us

Cooperative Extension

No specific information related to aquaculture from the state's Cooperative Extension Service was found. Complete information about the Illinois Cooperative Extension is available at:

http://www.extension.uiuc.edu

Sea Grant

The Illinois-Indiana Sea Grant College initiatives in aquaculture focus on the promotion and profitability of the industry in the two states. Outreach efforts include education, training, and promotion of research. Additional information about the Illinois-Indiana Sea Grant College can be found at:

http://www.iisgcp.org

INDIANA

NASAC

Paul Brown
Purdue University
1159 Forestry Building
West Lafayette IN 47907-1159
(765)494-4968
(765)496-2422 FAX
pb@fnr.purdue.edu

Cooperative Extension

No specific information related to aquaculture from the state's Cooperative Extension Service was found. Further information about the Purdue Extension can be found at:

http://www.ces.purdue.edu

Sea Grant

The Illinois-Indiana Sea Grant College initiatives in aquaculture focus on the promotion and profitability of the industry in the two states. Outreach efforts include education, training, and promotion of research. Additional information about the

Illinois-Indiana Sea Grant College can be found at:

http://www.iisgcp.org

IOWA

NASAC

Joe Morris
Department of Ecology
Iowa State University
124 Science II
Ames IA 50011
(515)294-4622
(515)294-7874 FAX
jemorris@ia.state.edu

Cooperative Extension

Iowa State University (ISU) Extension to Agricultural and Natural Resources provides information through its research and specialists about many topics including aquaculture. Links related to aquaculture include information about research initiatives and facilities and details about other organization and consortiums in the State or within the region. More information about ISU Extension to Agricultural and Natural Resources is available at:

http://www.extension.iastate.edu/ag/topics.html

Sea Grant

Iowa does not have a Sea Grant Program.

KANSAS

NASAC

Troy Amspacker
Kansas Department of Wildlife & Parks
Milford Fish Hatchery
3100 Hatchery Drive
Junction City KS 66411
(785)238 2638
(785)238-1369 FAX
troya@wp.state.ks.us

Cooperative Extension

K-State research provides information about aquaculture through its research specialists and publications. The K-State Extension lists aquaculture resources under their topic of farm ponds. The website has links to regional resources on aquaculture. Additional information about K-State research can be found at:

http://www.oznet.ksu.edu/root/coreResources.htm

Specific information about K-State Aquaculture is located at:

http://www.oznet.ksu.edu/neao/aquaculture.htm

Sea Grant

Kansas does not have a Sea Grant Program.

KENTUCKY

NASAC

Angela Caporelli
Kentucky Department of Agriculture
100 Fair Oaks Lane 5th floor
Frankfort KY 40601
(502)564-4983 ext 259
(502)564-0303 FAX
angela.caporelli@ky.gov

Cooperative Extension

The Cooperative Extension at the University of Kentucky makes available information relevant to aquaculture activities in the State. This information includes new reports of aquaculture projects in the State; details of events in different counties (e.g., shrimp farming events; and access to specialists. The Kentucky State University Aquaculture Program aids the extension through its research and specialists. Complete information about the Cooperative

Extension Service is located at:

http://www.ca.uky.edu/ces/index.htm

Further details of the Kentucky State University Aquaculture Program can be found at

http://www.ksuaquaculture.org

Sea Grant

Kentucky does not have a Sea Grant Program.

LOUISIANA

NASAC

Roy Johnson
Department of Agriculture and Forestry
PO Box 3334
Baton Rouge LA 70821-3334
225-922-1280
504-922-1289 FAX
roy_j@ldaf.state.la.us
www.ldaf.state.la.us

Cooperative Extension

The LSU AgCenter provides information about Cooperative Extension activities in Louisiana. Research and publications about aquaculture in Louisiana are available through the Center's website. Additional information about the LSU AgCenter is available at:

http://www.lsuagcenter.com

Sea Grant

The Louisiana Sea Grant, a participant in a 30-institution partnership, promotes responsible stewardship of marine and coastal resources through its program efforts. The Louisiana Sea Grant Program does not have a specific focus in aquaculture. Further information about the

Louisiana Sea Grant College can be obtained at:

http://www.laseagrant.org

MAINE

NASAC

John W. Sowles
Ecology Division
Maine Department of Marine Resources
P.O. Box 8
West Boothbay Harbor, ME 04575-0008
(207) 633-9518
(207) 633-9579 FAX
john.sowles@maine.gov

Cooperative Extension

The University of Maine Marine Extension Team is a cooperative effort between the University of Maine Cooperative Extension and the Maine Sea Grant College. The Marine Extension Team focuses on many aspects of aquaculture. The University of Maine Cooperative Extension can be accessed at:

http://www.umext.maine.edu

Sea Grant

The Maine Sea Grant College seeks to be a leader in marine science and education through its research and programming. A number of aquaculture programs continue to receive attention from the Maine Sea Grant. The Maine Sea Grant also cooperates with the Maine Marine Extension Team to conduct research and information sharing. Information about the Maine Sea Grant is located at:

http://www.seagrant.umaine.edu

MARYLAND

NASAC

Karl Roscher
Department of Agriculture
50 Harry S. Truman Parkway
Annapolis MD 21401-7080
(410)841-5724
(410)841-5970 FAX
roschekr@mda.state.md.us

Cooperative Extension

The University of Maryland Cooperative Extension Service does not focus specifically on aquaculture but does provide a wide range of information on other issues in resource management. Further information about the Cooperative Extension can be found at:

http://www.agnr.umd.edu/MCE/index.cfm

Sea Grant

The Maryland Sea Grant College focuses on marine research and education, with a concentration on the Chesapeake Bay, including aquaculture. Specialists at the Maryland Sea Grant Program are working to evaluate the economic efficiency of commercial finfish and shellfish aquaculture. Through Aquaculture Action, a teaching program, Sea Grant hosts workshops for educators to create a network of "aquaculture educators." Additional details about the Maryland Sea Grant Program can be found at:

http://www.mdsg.umd.edu/index.html

MASSACHUSETTS

NASAC

Scott Soares
Massachusetts Dept of Food and Agriculture
251 Causeway Street
Suite 500
Boston MA 02114-2151
617-626-1730
617-727-1850 FAX
scott.soares@state.ma.us
http://www.mass.gov/dfa/aquaculture

Cooperative Extension

The UMass Extension Service Division of Fish, Wildlife, and Biodiversity Conservation provides a variety of outreach and research and training efforts related to marine aquaculture. Further information about the UMass Extension is available at:

http://www.umass.edu/nrec

Sea Grant

WHOI Sea Grant provides an academic and research environment for addressing the revitalization of national fisheries and the development of sustainable aquaculture. The MIT Sea Grant provides educational resources on aquaculture through its K-12 curriculum at the finfish hatchery. WHOI Sea Grant information is located at:

http://www.whoi.edu/seagrant

Information on the MIT Sea Grant can be found at:

http://web.mit.edu/seagrant

MICHIGAN

NASAC

Nancy Frank
Michigan Department of Agriculture
Animal Industry Division
PO Box 30017
Lansing MI 48909
(517)373-1077
(517)373-6015 FAX
frankn@michigan.gov

Cooperative Extension

The Michigan State University Extension
has links to research publication focused on
aquaculture in the State and throughout the
Nation. Complete details of extension
programs can be found at:

http://www.msue.msu.edu/portal

Sea Grant

The Michigan Sea Grant promotes
protection and sustainable use of the aquatic
environment in the Great Lakes region.
Aquaculture is a major initiative of the Sea
Grant members in this region. Further
information about the Michigan Sea Grant
and its fisheries program can be found at:

http://www.miseagrant.umich.edu

MINNESOTA

NASAC

Ying Ji
Department of Agriculture
90 W Plato Boulevard
St Paul MN 55107
(651)296-5081
(651)296-6890 FAX
yingji@state.mn.us

Cooperative Extension

The University of Minnesota Extension
Service serves as a link between Minnesota

communities and the university. No specific
information related to aquaculture for the
state's Cooperative Extension Service was
found. Further information about the
University of Minnesota Extension Service
and publications about aquaculture can be
found at:

http://www.extension.umn.edu

Sea Grant

Minnesota Sea Grant uses research and
public education to further the state's coastal
environment and economy. Sea Grant acts
as a conduit for information among user
groups, including industry, management
agencies, and research scientists.
Aquaculture is a major outreach topic, and
information is available on beginning fish
farming operations and aquaculture business
structure. Additional information about
Minnesota Sea Grant is available at:

http://www.seagrant.umn.edu/index.html

Aquaculture outreach information can be
found at:

http://www.seagrant.umn.edu/aqua/index.html

MISSISSIPPI

NASAC

Gene Roberston
Department of Agriculture and Commerce
PO Box 1609
Jackson MS 39215-1609
(601)359-1102
(601)359-1174 FAX
gene@mdac.state.ms.us

Cooperative Extension

MSUcares (Coordinated Access to the
Research and Extension System) is a joint
effort between the Mississippi State
University Extension Service and the

Mississippi Agricultural and Forestry Experiment Station (MAFES). MSUcares provides information about aquaculture for residents of the state. More information about MSUcares is available at:

http://msucares.com

Sea Grant

The Mississippi-Alabama Sea Grant Consortium (MASGC) is an organization of nine universities and laboratories dedicated to activities that foster the conservation and sustainable development of coastal and marine resources in Mississippi and Alabama. Additional information about MASGC is available at:

http://www.masgc.org

MISSOURI

NASAC

Bart Hawcroft
Market Development Division
Missouri Department of Agriculture
P.O. Box 630
1616 Missouri Blvd
Jefferson City MO 65102
(573)526-6666
1-800-419-9139 (573)751-2868 FAX
bart.hawcroft@mda.mo.gov

Cooperative Extension

The Missouri Watershed Information Network's Aquaculture Center provides links to information pertaining to aquaculture in Missouri. Additional information about the Aquaculture Center can be found at:

http://outreach.missouri.edu/mowin/Training/aquaculture.html

Sea Grant
Missouri does not have a Sea Grant Program.

MONTANA

NASAC

Angie DeYoung
Department of Agriculture
Agriculture Development Division
PO Box 200201
Helena MT 59620-0201
(406)444-2402
(406)444-9442 FAX
adeyoung@state.mt.us
http://agr.state.mt.us/dept/agDevDiv.asp

Cooperative Extension

The Montana Cooperation Extension program is an educational resource with the aim of providing research-based knowledge to strengthen the social, economic and environmental well being of individuals, communities, and agricultural enterprises. This extension service does not have a particular focus on aquaculture. Further details of the Montana Cooperation Extension program can be found at:

http://extn.msu.montana.edu

Sea Grant

Montana does not have a Sea Grant Program.

NEBRASKA

NASAC

Ag Promotion and Development
Department of Agriculture
PO Box 94947
Lincoln NE 68509-4947
(402)471-4876
(402)471-2759 FAX
agprom@agr.state.ne.us

Cooperative Extension

The University of Nebraska Lincoln Cooperative Extension Service provides public information about current research efforts, offers training, and seeks to maintain educational resources about a variety of topics. Publications related to aquaculture are available at this site. More information about the University of Nebraska Lincoln Extension Service is located at:

http://www.extension.unl.edu

Publications may be searched at:

http://www.ianr.unl.edu/pubs

Sea Grant

Nebraska does not have a Sea Grant Program.

NEVADA

NASAC

No Program

Cooperative Extension

The University of Nevada Cooperative Extension provides local citizens with information for improving their lives and local surroundings. No specific information related to aquaculture for the state's Cooperative Extension Service was found. Additional information about the University of Nevada Cooperative Extension is available at:

http://www.unce.unr.edu

Sea Grant

Nevada does not have a Sea Grant Program.

NEW HAMPSHIRE

NASAC

J-J Newman (Joyce)
University of New Hampshire Sea Grant Division
Kingman Farm
Durham NH 03824
(603)749-1565
(603)743-3997 FAX
jj.newman@unh.edu

Cooperative Extension

UNH Aquaculture extension programs have the goals of assisting both potential and existing aquaculture operations with all aspects of the business including: species identification, broodstock care, nutrition, disease, systems design, marketing, permitting and business plans. Extension staff work closely with the aquaculture research community to highlight the most recent industry technologies. Further information of UNH Fisheries/Aquaculture programs is available at:

http://www.ceinfo.unh.edu

Sea Grant

The New Hampshire Sea Grant promotes research and education about aquatic resources. Sea Grant aquaculture extension programs are designed to aid potential and existing aquaculture operations in areas such as site and species evaluation, culture techniques, nutrition, health management, systems design, marketing, permitting, and business plans for a full range of finfish, shellfish, and seaweed operations. Complete information about the New Hampshire Sea Grant is available at:

http://www.seagrant.unh.edu/index.html

NEW JERSEY

NASAC

Linda O' Dierno
Fish and Seafood Development
Department of Agriculture CN 330
Trenton NJ 08625
(609)984-6757
(609)633-7229 FAX
linda.odierno@ag.state.nj.us

Cooperative Extension

The Rutgers Cooperative Extension aquaculture programs are focused on providing information about seafood productions in the state of New Jersey. Further details of the Rutgers Cooperative Extension program can be found at:

http://www.rce.rutgers.edu

Sea Grant

The New Jersey Sea Grant College, a part of the New Jersey Marine Sciences Consortium, provides training, research and educational materials focused on the marine environment in New Jersey. This Sea Grant has no specific programs in aquaculture. Further details of the New Jersey Sea Grant College are located at:

http://www.njmsc.org/Sea_Grant/About_SeaGrant.htm

NEW MEXICO

NASAC

Mike Sloane
New Mexico Department of Game & Fish
PO Box 25112
Sante Fe NM 87504
505-476-8055
505-476-8131 FAX
msloane@state.nm.us

Cooperative Extension

The Cooperative Extension at New Mexico State University provides educational outreach to New Mexico's citizens. Although no current programs in aquaculture exist, publications on this topic can be found at this website. Further information from New Mexico State University's Extension Service can be found at:

http://www.cahe.nmsu.edu/ces

Sea Grant

New Mexico does not have a Sea Grant program.

NEW YORK

NASAC

Philip Hulbert
Department of Environmental Conservation
50 Wolf Road Room 522
Albany NY 12233-4753
518-402-8920
518-485-5827 FAX
pxhulber@gw.dec.state.ny.us

Cooperative Extension

The Cornell Cooperative Extension Programs in Fish and Wildlife Biology provide information specific to fisheries management. Details of aquaculture programs are on a case-by-case basis. An example of a county-based aquaculture program is Suffolk County. Through its research and educational and training programs, the Suffolk County office of the Cooperative Extension seeks to improve the quality of aquaculture in this county of New York. More information about Cornell Cooperative Extension Programs can be found at:

http://www.dnr.cornell.edu/EXT/ext/fish&wildlife.htm

Suffolk Country information is available at:

http://www.cce.cornell.edu/counties/Suffolk/MARprograms/Aquaculturemain.htm

Sea Grant

The New York Sea Grant conducts research on a variety of topics related to marine ecosystems. Information about aquaculture and the New York Sea Grant was not available. Additional information about New York's Sea Grant program is available at:

http://www.seagrant.sunysb.edu

NORTH CAROLINA

NASAC

Debra Sloan
North Carolina Department of Agriculture
P.O. Box 1475
Franklin, NC 28744
(828)524-1264
(828)524-1264 FAX
debrasloan@earthlink.net
http://www.agr.state.nc.us/aquacult

Cooperative Extension

North Carolina State University (NCSU) North Carolina Agricultural and Technical State University (NC AT&T) Cooperative Extension provides information about ponds and aquaculture. NCSU/NC A&T Cooperative Extension is located at:

http://www.ces.ncsu.edu/copubs/ag/aqua

Sea Grant

The North Carolina Sea Grant College program directs research and information sharing about issues in coastal and marine resources. Aquaculture is one of the theme areas for the NC Sea Grant. The North

Carolina Sea Grant College Program can be accessed at:

http://www.ncsu.edu/seagrant

NORTH DAKOTA

NASAC

No Aquaculture Program

Cooperative Extension

The North Dakota State University (NDSU) Extension Service offers its citizens information and other educational resources. Aquaculture has become a topic of focus at the Carrington Research Extension, where economic development is the goal. Reports of the aquaculture program at the Carrington Research Extension are available at its website. Additional information about the NDSU Extension can be found at:

http://www.ext.nodak.edu

The Carrington Research Extension can be found at:

http://www.ag.ndsu.nodak.edu/carringt

Sea Grant

North Dakota does not have a Sea Grant Program.

OHIO

NASAC

Laura Tiu
Ohio State University Piketon Research Center
1864 Shyville Rd.
Piketon OH 45661-9749
740-289-2071
740-289-4591 FAX
tiu.2@osu.edu

Cooperative Extension

Providing aquaculture practitioners with information on topics such as fish culture methods, nutritional requirements, aquacultural system design and management, species selection and water quality management. Further information about the Ohio State Extension Aquaculture Program is available at:

http://piketon.osu.edu/aqua

Sea Grant

As a part of the Lake Erie Programs, the Ohio Sea Grant College offers publications and research initiatives focused on many topics in aquaculture. Additional information about the Ohio Sea Grant program can be found at:

http://www.sg.ohio-state.edu

OKLAHOMA

NASAC

Mitch Broiles
Oklahoma Department of Agriculture
2800 N Lincoln Boulevard
Oklahoma City OK 73105
405-522-6131
405-522-0756 FAX
mbroiles@oda.state.ok.us

Cooperative Extension

Oklahoma State University (OSU) Cooperative Extension's publications database, has links to a wide range of aquaculture documents. The Extension's Water Quality Team maintains a group of experts in the field of aquaculture and pond maintenance. Specifically, the OSU Extension's Southern District and expert Marley Beem maintain a site dedicated to aquaculture education.

Publications, contact information, and aquaculture resources from the Southern District are found at:

http://osuextra.okstate.edu,
http://biosystems.okstate.edu/waterquality/wqteam.htm, and
http://dasnr.okstate.edu/oces/sedistrict,
respectively.

Sea Grant

Oklahoma does not have a Sea Grant Program.

OREGON

NASAC

Dalton Hobbs
Oregon Department of Agriculture
Agriculture Development and Marketing Division
1207 NW Naito Parkway
Suite 104
Portland OR 97209-2832
503-872-6600
503-872-6601 FAX
dhobbs@oda.state.or.us

Cooperative Extension

No information concerning aquaculture for the state's cooperative extension service was available for Oregon. Additional information about Oregon State University's Extension Service can be found at:

http://extension.oregonstate.edu/index.php

Sea Grant

Research of the Oregon Sea Grant College is focused on promotion and sustainability of aquatic resources in the Northeast. Research in sustainable aquaculture is one initiative of this Sea Grant program. Complete information about the Oregon Sea Grant College can be obtained at:

http://seagrant.orst.edu

PENNSYLVANIA

NASAC

Kyle Nagurny
Agriculture Development
Department of Agriculture
2301 N Cameron Street
Harrisburg PA 17110-9408
(717)787-2376
(717)787-5643 FAX
knagurny@state.pa.us

Cooperative Extension

No information concerning aquaculture for
the state's Cooperative Extension Service
was available for Pennsylvania.

Sea Grant

The Pennsylvania Sea Grant has programs
dedicated to promoting research in
aquaculture in the Lake Erie region. Further
details of Pennsylvania Sea Grant programs
can be found at:

http://www.pserie.psu.edu/seagrant/seagindex.htm

PUERTO RICO

NASAC

Jaime Gonzalez Azar
Fisheries Development Program
Department of Agriculture
PO Box 10163
San Juan PR 00908-1163
(809)724-4911
(787)725-7884 FAX

Cooperative Extension

Although the University of Puerto Rico has
an Agricultural Extension Service, no
specific information related to aquaculture
was found. A current website was also
unavailable.

Sea Grant

The University of Puerto Rico Sea Grant
College Program is an educational program
devoted to the conservation and sustainable
use of coastal and marine resources in
Puerto Rico, the U.S. Virgin Islands and the
Caribbean region. Their mission is two-fold:
to conduct scientific research in the areas of
water quality, fisheries and mariculture,
seafood safety, marine recreation and coastal
tourism, coastal hazards and coastal
communities economic development; and to
apply their scientific knowledge to solve a
variety of problems that communities face
every day. For over two decades the
University of Puerto Rico Sea Grant College
Program has been working to promote
sustainable development and the wise use of
marine resources in Latin America and the
Caribbean region. Complete information
about the University of Puerto Rico Sea
Grant College Program is available at:

http://seagrant.uprm.edu

RHODE ISLAND

NASAC

Michelle Burnett
Rhode Island Department of Environmental
Management
Office of Marine Fisheries
3 Fort Wetherill Rd
Jamestown, RI 02835
401-423-1946
401-423-1925 FAX
michelle.burnett@dem.state.ri.us

Dave Alves
Oliver Stedman Government Center
4808 Tower Hill Rd
Wakefield RI 02879
401-783-3370
401-783-3767 FAX
dalves@crmc.state.ri.us

Cooperative Extension

The Rhode Island Cooperative Extension program in Aquaculture Biotechnology and Fishing seeks to increase local aquaculture through economically- and environmentally-beneficial technologies systems. The Rhode Island Cooperative Extension program in Aquaculture Biotechnology and Fishing website is:

http://www.uri.edu/ce/aquaculture.html

Sea Grant

The Rhode Island Sea Grant Sustainable Fisheries Extension Program seeks to revitalize and stabilize the Nation's Fisheries through applied research, outreach, and education. The program provides a link between researchers, commercial and recreational fishermen, and regulators by bringing scientific and technical information to user groups, and in turn informing researchers of user needs and priorities. Fisheries Extension deals with capture fisheries, aquaculture, and seafood safety and quality. Information about the Rhode Island Sea Grant can be found at:

http://seagrant.gso.uri.edu

SOUTH CAROLINA

NASAC

Gerry Bonnette
South Carolina Department of Agriculture
PO Box 11280
Columbia SC 29211-1280
803-734-2218
803-734-0325 FAX
gbonnett@scda.sc.gov

Cooperative Extension

Clemson's Aquacultural Water Resources website provides links to a variety of information sources including publications, meetings, contacts, and research. Links to Clemson Aquacultural Water Resources can be found at:

http://www.clemson.edu/waterquality/waterres/aqwr.htm

Sea Grant

The South Carolina Sea Grant Consortium provides a program of research, education, extension, and training to increase economic opportunities and conservation of coastal and marine resources for citizens of South Carolina. Complete information about the SC Sea Grant Consortium can be found at:

http://www.scseagrant.org

SOUTH DAKOTA

NASAC

Jon Farris
South Dakota Department of Agriculture
Division of Agricultural Development
Foss Building
523 E Capitol
Pierre SD 57501-3182
(605)773-5436
(605)773-3481 FAX
jon.farris@state.sd.us

Dennis Unkenholz
South Dakota Department of Game, Fish and Parks
521 E Capitol
Pierre SD 57501
(605)773-4508
(605)773-6245 FAX
dennis.unkenholz@state.sd.us

Cooperative Extension

South Dakota State University's Extension Service is the primary outreach facility of the University. It serves the people of South Dakota by helping them apply scientific knowledge to improve their lives. No information on aquaculture-related programs is available for this Extension. Additional information about South Dakota State University's Cooperative Extension is located at:

http://sdces.sdstate.edu

Sea Grant

There is no Sea Grant Program for the State of South Dakota.

TENNESSEE

NASAC

Robert Beets
Tennessee Department of Agriculture
Market Development Division
P O Box 40627
Nashville TN 37204
615-837 5517
615-837 5194 FAX
robert.beets@state.tn.us

Cooperative Extension

The Center for Profitable Agriculture is a partnership between the Tennessee Farm Bureau and the University of Tennessee Institute of Agriculture. The Center provides information about the development of the aquaculture industry in Tennessee. Further information about aquaculture initiatives from the Center for Profitable Agriculture can be found on the University of Tennessee Extension's web site at:

http://www.utextension.utk.edu/departments

Sea Grant

There is no Sea Grant Program in Tennessee.

TEXAS

NASAC

Susan Dunn
Institutional and Produce Marketing
Texas Department of Agriculture
PO Box 12847
Austin TX 78711
(512)475-1665
(512)463-7843 FAX
susan.dunn@agr.state.tx.us

Bob Blumberg
General Land Office
1700 N Congress Avenue
SFA Building Room 710
Austin TX 78701-1495
(512)463-5028
(512)463-5098 FAX

Cooperative Extension

The Texas A&M Cooperative Extension offers practical, educational information as a result of university research. No specific information related to aquaculture from the state's Cooperative Extension Service was found. Texas A&M Cooperative Extension information can be found at:

http://texasextension.tamu.edu

Sea Grant

Texas Sea Grant provides outreach through two efforts: the Marine Advisory Service (MAS) and the Marine Information Service (MIS). The Texas Sea Grant has specialists who deal with issues including aquaculture, fisheries, environmental quality, marine business management, marine education, and seafood science, technology and marine policy. Further information about the Texas Sea Grant College Program can be found at:

http://texas-sea-grant.tamu.edu

UTAH

NASAC

Kent Hauck
Utah Department of Agriculture & Food
P.O. Box 146500
Salt Lake City UT 84114-6500
(801)538-7029
(801)538-7169 FAX
khauck@utah.gov

Cooperative Extension

No specific information related to aquaculture from the state's Cooperative Extension Service was found. Further information about initiatives from the Utah State University Extension Service is located at:

http://extension.usu.edu

Sea Grant

Utah does not have a Sea Grant Program.

VERMONT

NASAC

Denise Russo
Vermont Department of Agriculture
116 State Street Drawer 20
Montpelier VT 05620-2901
(802)828-3829
(802)828-3831 FAX
drusso@agr.state.vt.us

Cooperative Extension

The University of Vermont's Extension seeks to help build vital, sustainable rural communities and families through educational programs. Additional information about the University of Vermont Extension is available at:

http://www.uvm.edu/extension

Sea Grant

The Lake Champlain Sea Grant Program provides a variety of scientific research and activities that lead to improved understanding, use, and management of the Lake Champlain ecosystem. Information about aquaculture through the Lake Champlain Sea Grant was not found. Additional information about the Lake Champlain Sea Grant is available at:

http://www.uvm.edu/~seagrant

VIRGINIA

NASAC

T. Robins Buck
Department of Agriculture and Consumer Services
1100 Bank Street Suite 210
Richmond VA 23219
(804)371-6094
(804)371-2945 FAX
robins.buck@vdacs.virginia.gov

Cooperative Extension

The Southwest Virginia Aquaculture Research and Extension Center seeks to aid the public with sustainable recirculating aquaculture and high value alternative horticulture opportunities in southwest Virginia through research, extension, and education programs. More information from the Southwest Virginia Aquaculture Research and Extension Center is located at:

http://www.vaes.vt.edu/saltville

Sea Grant

The Virginia Sea Grant Program is a consortium of research organizations that focuses on protection and use of marine and freshwater resources in the State. Development of sustainable aquaculture is a key research initiative from this program.

More information about the Virginia Sea Grant is available at:

http://www.virginia.edu/virginia-sea-grant

VIRGIN ISLANDS

NASAC

Arthur Petersen, Jr.
Virgin Islands Department of Agriculture
Estate Lower Love
Kingshill, St. Croix VI 00850
(340)778-0991
(340)778-3101

Cooperative Extension

The University of the Virgin Islands (UVI) Cooperative Extension Service, Agricultural Experiment Station (AES) is located on the St. Croix Campus of the University of the Virgin Islands. AES is part of the Research and Public Service Component.

AES conducts basic and applied research to meet the needs of the local agricultural community in increasing production, improving efficiency, developing new enterprises, preserving and propagating germplasm unique to the Virgin Islands, and protecting the natural resource base. AES has a research program for aquaculture.

Information about the UVI Cooperative Extension Service is available at:

http://rps.uvi.edu/CES

Additional information about AES is available at:

http://rps.uvi.edu/AES/aes_home.html

Sea Grant

The Virgin Islands Marine Advisory Service (VIMAS), a part of the University of Puerto

Rico Sea Grant College Program, is located within the Center for Marine and Environmental Studies (CMES) at UVI. VIMAS was established on the St. Thomas campus of UVI in 1984 and later expanded to include agents on St. Croix. No specific information related to aquaculture for the Virgin Islands' Cooperative Extension Service was found. Additional information about VIMAS is available at:

http://rps.uvi.edu/VIMAS

WASHINGTON

NASAC

Dan Swecker, Senator
Washington State Aquaculture Coordinator
10420 173rd Ave. SW
Rochester WA 98579
360-273-5890
360-273-6577 FAX
dan@wfga.net

Cooperative Extension

Washington State University Cooperative Extension offers non-credit education and degree opportunities to people throughout the state. Cooperative Extension builds the capacity of individuals, organizations, businesses and communities, empowering them to find solutions for local issues and to improve their quality of life. WSU Cooperative Extension offers some aquaculture publications at:

http://pubs.wsu.edu/cgi-bin/pubs/index.html

Additional information about the WSU Cooperative Extension is available at:

http://ext.wsu.edu

Sea Grant

Washington Sea Grant Program brings researchers, regulatory agencies and industry together to keep Washington-raised aquatic species available for sale as food products. Through research, education, and information sharing, Washington State aquaculture continues to be productive. Further information about the Washington Sea Grant Program can be found at:

http://www.wsg.washington.edu

WEST VIRGINIA

NASAC

Rob Nichols
West Virginia Department of Agriculture
1900 Kanawha Boulevard, East
Charleston WV 25305
(304)558-2208
(304)558-3594 FAX
rnichols@ag.state.wv.us

Cooperative Extension

The WVU Extension Service has established a website dedicated to promoting education, technical support, and cooperative resources related to the aquaculture industry. Additional resources from the WVU Extension Service are available at:

http://www.wvu.edu/~agexten/aquaculture

Sea Grant

West Virginia does not have a Sea Grant Program.

WISCONSIN

NASAC

Will H. Hughes
Wisconsin Department of Agriculture, Trade & Consumer Protection
Division of Agricultural Development
2811 Agriculture Drive
Madison, WI 53708-8911
608-224-5142
Fax: 608-224-5110
will.hughes@datcp.state.wi.us

Cooperative Extension

Through its programming and collaborative relationships with the UW universities and colleges, the 72 Wisconsin counties, and countless local, state, and federal agencies and groups, Extension provides a spectrum of lifelong learning opportunities for Wisconsin citizens. Extension education applies university research, knowledge and resources to the needs of Wisconsin citizens. Aquaculture is a topic of interest for the Wisconsin Extension. Further information about University of Wisconsin's Extension can be found at:

http://www.uwex.edu

Sea Grant

The Wisconsin Sea Grant provides information on research and publications pertaining to aquaculture. Through these efforts as well as initiatives in technology transfer and outreach, Wisconsin Sea Grant acts to promote sustainable use of aquatic resources in the Great Lakes and the Oceans. Additional information about Wisconsin's Sea Grant Programs can be found at:

http://www.seagrant.wisc.edu/index.asp

WYOMING

NASAC

Wyoming Business Council
Division of Agriculture and Timber
2219 Carey Avenue
Cheyenne WY 82002-0100
(307)777-6577
(307)777-6593 FAX
wda@state.wy.us

Cooperative Extension

The University of Wyoming Cooperative
Extension offers access to information and
research on a variety of topics related to
community life in the State. Additional
information about aquaculture through this
Extension was not found. The University's
Extension can be accessed at:

http://uwadmnweb.uwyo.edu/UWces

Sea Grant

Wyoming does not have a Sea Grant
Program.

Appendix I

Additional Resources

Additional Resources

The following resources may provide useful information to owners and operators of aquatic animal production facilities.

EPA Programs and Information

CAAP Final Rule Web Page

This website provides access to the text of the rule and preamble, supporting/guidance documents.
http://epa.gov/guide/aquaculture

 **U.S. EPA NPDES Permit Writers' Manual, EPA 833-B-96-003, December 1, 1996.**

You may download individual chapters or the entire document at:
http://cfpub.epa.gov/npdes/writermanual.cfm?prog

NPDES Permit Program Basics

This Web site provides basic permitting tools and information.
http://cfpub.epa.gov/npdes/home.cfm?program_id=45

Permit Compliance System

http://www.epa.gov/enviro/html/pcs/pcs_overview.html#PCS

Source Water Protection Programs

EPA Office of Groundwater and Drinking Water, Source Water Protection.
http://www.epa.gov/safewater/protect.html

TMDL Programs

EPA Office of Wetlands, Oceans and Watersheds, TMDL Program.
http://www.epa.gov/OWOW/tmdl/index.html

USDA Programs and Information

Land Grant Universities

This website provides directory of land grant universities. Click on a state link to reach a list of land grant university web sites.

http://www.csrees.usda.gov/qlinks/partners/state_partners.html

USDA Agricultural Research Service

http://www.ars.usda.gov

USDA Animal and Plant Health Inspection Service (APHIS)

http://www.aphis.usda.gov

USDA Cooperative State Research, Education, and Extension Service (CSREES)

http://www.csrees.usda.gov

USDA National Agricultural Statistics Service (NASS)

http://www.nass.usda.gov/index.asp

USDA Natural Resources Conservation Service (NRCS)

http://www.nrcs.usda.gov

USDA NRCS Conservation Programs

Environmental Quality Incentives Program, Agricultural Management Assistance Program, Wetlands Reserve Program, Wildlife Habitat Incentives Program.
http://www.nrcs.usda.gov/programs

USDA Regional Aquaculture Centers

- **Center for Tropical and Subtropical Aquaculture:** http://www.ctsa.org
- **North Central Regional Aquaculture Center:** http://www.ncrac.org/
- **Northeastern Regional Aquaculture Center:** http://www.agnr.umd.edu/AGNRDirectory/Section.cfm?SN=%208.13
- **Southern Regional Aquaculture Center:** http://www.msstate.edu/dept/srac
- **Western Regional Aquaculture Center:** http://www.fish.washington.edu/wrac

Other Resources

American Fisheries Society
http://www.fisheries.org

American Tilapia Association
http://ag.arizona.edu/azaqua/ata.html

AquaFeed.com
http://www.aquafeed.com

Aquaculture Engineering Society
http://www.aesweb.org

Aquaculture Magazine
http://www.aquaculturemag.com

Aquaculture Network Information Center (Aquanic)
http://www.aquanic.org

Atlantic Salmon Federation
http://www.asf.ca

Controlling Birds at Aquaculture Facilities
http://pubs.cas.psu.edu/FreePubs/pdfs/uh120.pdf

Freshwater Institute
http://www.freshwaterinstitute.org

Global Aquaculture Alliance
http://www.gaalliance.org

Joint Subcommittee on Aquaculture
http://ag.ansc.purdue.edu/aquanic/jsa/index.htm

Maryland Department of Agriculture, Directory of State Aquaculture Coordinators and Contacts
http://www.marylandseafood.org/aquaculture/nasac.php

National Aquaculture Association
http://www.nationalaquaculture.org

National Association of State Aquaculture Coordinators
http://www.agr.state.nc.us/aquacult/NASAC.html

National Fisheries Institute
http://www.nfi.org

National Marine Fisheries Service
http://www.nmfs.noaa.gov

National Sea Grant Library
http://nsgl.gso.uri.edu/index.html

NOAA Aquaculture Information Center
http://www.lib.noaa.gov/docaqua/frontpage.htm

Pacific Coast Shellfish Growers Association
http://www.pcsga.org

Pacific Shellfish Institute
http://www.pacshell.org

Sea Grant
http://www.nsgo.seagrant.org

SeaWeb Aquaculture Clearinghouse
http://www.seaweb.org/resources/sac

Southern Regional Aquaculture Center Fact Sheets
http://srac.tamu.edu

U.S. Army Corps of Engineers
http://www.usace.army.mil

U.S. Fish and Wildlife Service
http://www.fws.gov

U.S. Food and Drug Administration, Center for Veterinary Medicine (CVM)
http://www.fda.gov/cvm/default.html

U.S. Trout Farmers Association
http://www.ustfa.org

Appendix J

Glossary

Glossary

Aeration: The process of bringing air into contact with a liquid by one or more of the following methods: (1) spraying the liquid into the air, (2) bubbling air through the liquid, and (3) agitating the liquid to promote absorption of oxygen through the air-liquid interface.

Aerobic: Having or occurring in the presence of free oxygen.

Agronomic rates: The land application of animal wastes at rates of application that provide the crop or forage growth with needed nutrients for optimum health and growth.

Anaerobic: Characterized by the absence of molecular oxygen, or capable of living and growing in the absence of oxygen, such as *anaerobic bacteria.*

Aquaculture: The propagation and rearing of aquatic species in controlled or selected environments.

Aquatic animal production: The production of aquatic animals under controlled or semicontrolled conditions.

Benthic monitoring: Monitoring conducted to ensure that degradation is not occurring under or around net pens.

Best Available Technology Economically Achievable (BAT): Technology-based standard established by the Clean Water Act (CWA) as the most appropriate means available on a national basis for controlling the direct discharge of toxic and nonconventional pollutants to navigable waters. BAT effluent limitations guidelines, in general, represent the best existing performance of treatment technologies that are economically achievable within an industrial point source category or subcategory.

Best Conventional Pollutant Control Technology (BCT): Technology-based standard for the discharge from existing industrial point sources of conventional pollutants including BOD, TSS, fecal coliform, pH, oil and grease. The BCT is established in light of a two-part "cost reasonableness" test, which compares the cost for an industry to reduce its pollutant discharge with the cost to a POTW for similar levels of reduction of a pollutant loading. The second test examines the cost-effectiveness of additional industrial treatment beyond BPT. EPA must find limits, which are reasonable under both tests before establishing them as BCT.

Best management practices: Schedules of activities, prohibitions of practices, maintenance procedures, and other management practices that prevent or reduce pollution (Title 40 CFR Part 122.2).

Best Practicable Control Technology Currently Available (BPT): The first level of technology-based standards established by the CWA to control pollutants discharged to waters of the United States. BPT effluent limitations guidelines are generally based on the average of the best existing performance by plants within an industrial category or subcategory.

Biosolids: Waste material from an aquaculture operation, primarily fish manure and uneaten feed.

Clean Water Act (CWA): The Clean Water Act is an act passed by the U.S. Congress to control water pollution. It was formerly referred to as the Federal Water Pollution Control Act of 1972 or Federal Water Pollution Control Act Amendments of 1972 (Public Law 92-500), 33 U.S.C. 1251 et. seq., as amended by: Public Law 96-483; Public Law 97- 117; Public Laws 95-217, 97-117, 97-440, and 100-04.

Concentrated aquatic animal production (CAAP) facility: A hatchery, fish farm, or other facility that contains, grows, or holds aquatic animals in either of the following categories, or that the Director designates as such on a case-by-case basis, and must apply for a National Pollutant Discharge Elimination System permit.

Coldwater fish species or other coldwater aquatic animals including, but not limited to, the Salmonidae family of fish (e.g., trout and salmon) in ponds, raceways, or other similar structures that discharge at least 30 days per year but does not include:

(1) Facilities that produce less than 9,090 harvest weight kilograms (approximately 20,000 pounds) of aquatic animals per year and
(2) Facilities that feed less than 2,272 kilograms (approximately 5,000 pounds) of food during the calendar month of maximum feeding.

Warmwater fish species or other warmwater aquatic animals including, but not limited to, the Ameiuridae, Cetrachidae, and the Cyprinidae families of fish (e.g., respectively, catfish, sunfish, and minnows) in ponds, raceways, or similar structures that discharge at least 30 days per year, but does not include:

(1) Closed ponds that discharge only during periods of excess runoff or
(2) Facilities that produce less than 45,454 harvest weight kilograms (approximately 100,000 pounds) of aquatic animals per year.

Drug: Any substance, including medicated feed, that is added to a production facility to maintain or restore animal health and that subsequently might be discharged to waters of the United States.

Effluent limitations guidelines (ELGs): Under the Clean Water Act, section 502(11), any restriction, including schedules of compliance, established by a state or the Administrator on quantities, rates, and concentrations of chemical, physical, biological, and other constituents that are discharged from point sources into navigable waters, the waters of the contiguous zone, or the ocean (Clean Water Act sections 301(b) and 304(b)).

Excess feed: Feed that is added to a production system, is not consumed, and is not expected to be consumed by the aquatic animals.

Existing source: Any facility from which there is or may be a discharge of pollutants, the construction of which is commenced before September 22, 2004.

Extralabel use: The use of a drug in any way that is not in accordance with approved labeling. Extralabel use may be allowed under specific conditions.

Facility: All contiguous property and equipment owned, operated, leased, or under the control of the same person or entity.

Feed conversion ratio (FCR): A measure of feeding efficiency that is calculated as the ratio of the weight of feed applied to the weight of the fish produced.

Flow-through systems: A system designed for a continuous water flow to waters of the United States through chambers used to produce aquatic animals. Flow-through systems typically use either raceways or tank systems. Raceways are fed by nearby rivers or springs and are typically long, rectangular chambers at or below grade, constructed of earth, concrete, plastic, or metal. Tank systems are similarly fed and concentrate aquatic animals in circular or rectangular tanks above grade. The term does not include net pens.

Groundwater: Water in a saturated zone or stratum beneath the surface of land or water.

Indirect discharger: A facility that discharges or may discharge wastewaters into a publicly owned treatment works.

National Pollutant Discharge Elimination System (NPDES) permit: A permit to discharge wastewater into waters of the United States issued under the National Pollutant Discharge Elimination System, authorized by section 402 of the Clean Water Act.

National Pollutant Discharge Elimination System (NPDES) program: The NPDES program authorized by sections 307, 318, 402, and 405 of the Clean Water Act. It applies to facilities that discharge wastewater directly to U.S. surface waters.

Navigable waters: Traditionally, waters sufficiently deep and wide for navigation by all, or specified vessels; such waters in the United States come under federal jurisdiction and are protected by certain provisions of the Clean Water Act.
Net pens and cage systems: A culture system that uses suspended or floating systems to culture fish or shellfish. These systems may be located along a shore or pier or may be anchored and floating offshore. Net pens and cages rely on tides, currents, and other natural water movement to provide a continual supply of high-quality water to the cultured animals.

New Source Performance Standards (NSPS): Technology-based standards for facilities that qualify as new sources under 40 CFR 122.2 and 40 CFR 122.29. Standards consider that the new source facility has an opportunity to design operations to more effectively control pollutant discharges.

Outfall: The mouth of the conduit drains and other conduits from which a facility effluent discharges into receiving waters.

Pass through: A discharge which exits the POTW into waters of the United States, or state of Washington, in quantities or concentrations which, alone or in conjunction with a discharge or discharges from other sources, is a cause of a violation of any requirement of the city's NPDES permit including an increase in the magnitude or duration of a violation.

Permitting authority: The agency authorized to administer the National Pollutant Discharge Elimination System permitting program in a state or territory.

Point source: Any discernible, confined, and discrete conveyance from which pollutants are or may be discharged. See Clean Water Act section 502(14).

Ponds: Culture systems characterized by hydraulic retention times sufficiently long to allow natural processes to reduce metabolic waste concentrations. Commonly used to culture warm water fish, such as channel catfish.

Pretreatment standards for existing sources (PSES) of indirect discharges: Under section 307(b) of the Clean Water Act, standards applicable (for this rule) to indirect dischargers that commenced construction prior to promulgation of the final rule.

Pretreatment standards for new sources (PSNS): Under section 307(c) of the Clean Water Act, standards applicable to indirect dischargers that commence after promulgation of the final rule.

Publicly owned treatment works (POTW): A treatment works as defined by section 212 of the Clean Water Act, which is owned by a state or municipality (as defined by section 502(4) of the Clean Water Act). This definition includes any devices and systems used in the storage, treatment, recycling, and reclamation of municipal sewage or industrial wastes of a liquid nature. It also includes sewers, pipes, and other conveyances, only if they convey wastewater to a POTW. The term also means the municipality, as defined in section 502(4) of the Clean Water Act, that has jurisdiction over the indirect discharges to and the discharges from such a treatment works.

Quiescent zones: Solids-collection zones placed at the end of a raceway tank to collect the settleable solids swept out of the fish-rearing area. They are the primary means for solids removal in flow-through raceways.

Raceways: Culture units in which water flows continuously, making a single pass through the unit before being discharged; these systems are also referred to as flow-through systems.

Recirculating systems: A system that filters and reuses water in which aquatic animals are produced prior to discharge. Recirculating systems typically use tanks, biological or mechanical filtration, and mechanical support equipment to maintain high-quality water to produce aquatic animals. These systems are highly intensive and require biological treatment within the system to prevent ammonia from accumulating to harmful levels.

Sludge: Settled sewage solids combined with varying amounts of water and dissolved materials that are removed from sewage by screening, sedimentation, chemical precipitation, or bacterial digestion.

Wastewater treatment: The processing of wastewater by physical, chemical, biological, or other means to remove specific pollutants from the wastewater stream, or to alter the physical or chemical state of specific pollutants in the wastewater stream. Treatment is performed for discharge of treated wastewater, recycle of treated wastewater to the same process that generated the wastewater, or reuse of the treated wastewater in another process.

Appendix K1

NPDES Permit Applications: Form 1

Please print or type in the unshaded areas only
(fill—in areas are spaced for elite type, i.e., 12 characters/inch).

Form Approved. OMB No. 2040-0086.

FORM 1 GENERAL	⊕EPA	U.S. ENVIRONMENTAL PROTECTION AGENCY **GENERAL INFORMATION** *Consolidated Permits Program* *(Read the "General Instructions" before starting.)*	I. EPA I.D. NUMBER

I. EPA I.D. NUMBER

F

LABEL ITEMS

I. EPA I.D. NUMBER

III. FACILITY NAME

V. FACILITY MAILING ADDRESS

VI. FACILITY LOCATION

PLEASE PLACE LABEL IN THIS SPACE

GENERAL INSTRUCTIONS

If a preprinted label has been provided, affix it in the designated space. Review the information carefully; if any of it is incorrect, cross through it and enter the correct data in the appropriate fill—in area below. Also, if any of the preprinted data is absent *(the area to the left of the label space lists the information that should appear)*, please provide it in the proper fill—in area(s) below. If the label is complete and correct, you need not complete Items I, III, V, and VI *(except VI-B which must be completed regardless)*. Complete all items if no label has been provided. Refer to the instructions for detailed item descriptions and for the legal authorizations under which this data is collected.

II. POLLUTANT CHARACTERISTICS

INSTRUCTIONS: Complete A through J to determine whether you need to submit any permit application forms to the EPA. If you answer "yes" to any questions, you must submit this form and the supplemental form listed in the parenthesis following the question. Mark "X" in the box in the third column if the supplemental form is attached. If you answer "no" to each question, you need not submit any of these forms. You may answer "no" if your activity is excluded from permit requirements; see Section C of the instructions. See also, Section D of the instructions for definitions of bold—faced terms.

SPECIFIC QUESTIONS	YES	NO	FORM ATTACHED	SPECIFIC QUESTIONS	YES	NO	FORM ATTACHED
A. Is this facility a **publicly owned treatment works** which results in a **discharge to waters of the U.S.?** (FORM 2A)	16	17	18	B. Does or will this facility *(either existing or proposed)* include a **concentrated animal feeding operation** or **aquatic animal production facility** which results in a **discharge to waters of the U.S.?** (FORM 2B)	19	20	21
C. Is this a facility which currently results in **discharges** to **waters of the U.S.** other than those described in A or B above? (FORM 2C)	22	23	24	D. Is this a proposed facility *(other than those described in A or B above)* which will result in a **discharge** to **waters of the U.S.?** (FORM 2D)	25	26	27
E. Does or will this facility treat, store, or dispose of **hazardous wastes?** (FORM 3)	28	29	30	F. Do you or will you inject at this facility industrial or municipal effluent below the lowermost stratum containing, within one quarter mile of the well bore, underground sources of drinking water? (FORM 4)	31	32	33
G. Do you or will you inject at this facility any produced water or other fluids which are brought to the surface in connection with conventional oil or natural gas production, inject fluids used for enhanced recovery of oil or natural gas, or inject fluids for storage of liquid hydrocarbons? (FORM 4)	34	35	36	H. Do you or will you inject at this facility fluids for special processes such as mining of sulfur by the Frasch process, solution mining of minerals, in situ combustion of fossil fuel, or recovery of geothermal energy? (FORM 4)	37	38	39
I. Is this facility a proposed **stationary source** which is one of the 28 industrial categories listed in the instructions and which will potentially emit 100 tons per year of any air pollutant regulated under the Clean Air Act and may affect or be located in an **attainment area?** (FORM 5)	40	41	42	J. Is this facility a proposed **stationary source** which is NOT one of the 28 industrial categories listed in the instructions and which will potentially emit 250 tons per year of any air pollutant regulated under the Clean Air Act and may affect or be located in an **attainment area?** (FORM 5)	43	44	45

III. NAME OF FACILITY

C 1 | SKIP

IV. FACILITY CONTACT

A. NAME & TITLE *(last, first, & title)* | B. PHONE *(area code & no.)*

C 2

V. FACILITY MAILING ADDRESS

A. STREET OR P.O. BOX

C 3

B. CITY OR TOWN | C. STATE | D. ZIP CODE

C 4

VI. FACILITY LOCATION

A. STREET, ROUTE NO. OR OTHER SPECIFIC IDENTIFIER

C 5

B. COUNTY NAME

C. CITY OR TOWN | D. STATE | E. ZIP CODE | F. COUNTY CODE *(if known)*

C 6

EPA Form 3510-1 (8-90)

CONTINUE ON REVERSE

CONTINUED FROM THE FRONT

VII. SIC CODES (4-digit, in order of priority)

A. FIRST	B. SECOND
c 7 (specify) 15 16 - 19	c 7 (specify) 11 15 - 19
C. THIRD	D. FOURTH
c 7 (specify) 15 16 - 19	c 7 (specify) 15 16 - 19

VIII. OPERATOR INFORMATION

A. NAME	B. Is the name listed in Item VIII-A also the owner?
c 8 15 16 55	☐ YES ☐ NO 66

C. STATUS OF OPERATOR (Enter the appropriate letter into the answer box; if "Other", specify.)

F = FEDERAL   M = PUBLIC (other than federal or state)   (specify)
S = STATE   O = OTHER (specify)
P = PRIVATE   56

D. PHONE (area code & no.)

c A 15   16 - 18   19 - 21   22 - 25

E. STREET OR P.O. BOX

26   55

F. CITY OR TOWN	G. STATE	H. ZIP CODE	IX. INDIAN LAND
c B 15 16	40 41 42	47 - 51	Is the facility located on Indian lands? ☐ YES ☐ NO 52

X. EXISTING ENVIRONMENTAL PERMITS

A. NPDES (Discharges to Surface Water)	D. PSD (Air Emissions from Proposed Sources)	
c T I 9 N 15 16 17 18 30	c T I 9 P 15 16 17 18 30	
B. UIC (Underground Injection of Fluids)	E. OTHER (specify)	(specify)
c T I 9 U 15 16 17 18 30	c T I 9 15 16 17 18 30	
C. RCRA (Hazardous Wastes)	E. OTHER (specify)	(specify)
c T I 9 R 15 16 17 18 30	c T I 9 15 16 17 18 30	

XI. MAP

Attach to this application a topographic map of the area extending to at least one mile beyond property bounderies. The map must show the outline of the facility, the location of each of its existing and proposed intake and discharge structures, each of its hazardous waste treatment, storage, or disposal facilities, and each well where it injects fluids underground. Include all springs, rivers and other surface water bodies in the map area. See instructions for precise requirements.

XII. NATURE OF BUSINESS (provide a brief description)

XIII. CERTIFICATION (see instructions)

I certify under penalty of law that I have personally examined and am familiar with the information submitted in this application and all attachments and that, based on my inquiry of those persons immediately responsible for obtaining the information contained in the application, I believe that the information is true, accurate and complete. I am aware that there are significant penalties for submitting false information, including the possibility of fine and imprisonment.

A. NAME & OFFICIAL TITLE (type or print)	B. SIGNATURE	C. DATE SIGNED

COMMENTS FOR OFFICIAL USE ONLY

c C 15 16   55

EPA Form 3510-1 (8-90)

United States
Environmental Protection
Agency

Office of
Enforcement
Washington, DC 20460

EPA Form 3510-1
Revised August 1990

Permits Division

Application Form 1 - General Information

Consolidated Permits Program

This form must be completed by all persons applying for a permit under EPA's Consolidated Permits Program. See the general instructions to Form 1 to determine which other application forms you will need.

DESCRIPTION OF CONSOLIDATED PERMIT APPLICATION FORMS

The Consolidated Permit Application Forms are:

Form 1 — General Information *(included in this part)*;

Form 2 — Discharges to Surface Water *(NPDES Permits)*:

2A. Publicly Owned Treatment Works *(Reserved — not included in this package)*,

2B. Concentrated Animal Feeding Operations and Aquatic Animal Production Facilities *(not included in this package)*,

2C. Existing Manufacturing, Commercial, Mining, and Silvicultural Operations *(not included in this package)*, and

2D. New Manufacturing, Commercial, Mining, and Silvicultural Operations *(Reserved — not included in this package)*;

Form 3 — Hazardous Waste Application Form *(RCRA Permits — not included in this package)*;

Form 4 — Underground Injection of Fluids *(UIC Permits — Reserved — not included in this package)*; and

Form 5 — Air Emissions in Attainment Areas *(PSD Permits — Reserved — not included in this package)*.

SECTION A — GENERAL INSTRUCTIONS

Who Must Apply

With the exceptions described in Section C of these instructions, Federal laws prohibit you from conducting any of the following activities without a permit.

NPDES *(National Pollutant Discharge Elimination System Under the Clean Water Act, 33 U.S.C. 1251)*. Discharge of pollutants into the waters of the United States.

RCRA *(Resource Conservation and Recovery Act, 42 U.S.C. 6901)*. Treatment, storage, or disposal of hazardous wastes.

UIC *(Underground Injection Control Under the Safe Drinking Water Act, 42 U.S.C. 300f)*. Injection of fluids underground by gravity flow or pumping.

PSD *(Prevention of Significant Deterioration Under the Clean Air Act, 72 U.S.C. 7401)*. Emission of an air pollutant by a new or modified facility in or near an area which has attained the National Ambient Air Quality Standards for that pollutant.

Each of the above permit programs is operated in any particular State by either the United States Environmental Protection Agency *(EPA)* or by an approved State agency. You must use this application form to apply for a permit for those programs administered by EPA. For those programs administered by approved States, contact the State environmental agency for the proper forms.

If you have any questions about whether you need a permit under any of the above programs, or if you need information as to whether a particular program is administered by EPA or a State agency, or if you need to obtain application forms, contact your EPA Regional office *(listed in Table 1)*.

Upon your request, and based upon information supplied by you, EPA will determine whether you are required to obtain a permit for a particular facility. Be sure to contact EPA if you have a question, because Federal laws provide that **you may be heavily penalized if you do not apply for a permit when a permit is required.**

Form 1 of the EPA consolidated application forms collects general information applying to all programs. You must fill out Form 1 regardless of which permit you are applying for. In addition, you must fill out one of the supplementary forms *(Forms 2 — 5)* for each permit needed under each of the above programs. Item II of Form 1 will guide you to the appropriate supplementary forms.

You should note that there are certain exclusions to the permit requirements listed above. The exclusions are described in detail in Section C of these instructions. If your activities are excluded from permit requirements then you do not need to complete and return any forms.

NOTE: Certain activities not listed above also are subject to EPA administered environmental permit requirements. These include permits for ocean dumping, dredged or fill material discharging, and certain types of air emissions. Contact your EPA Regional office for further information.

Table 1. Addresses of EPA Regional Contacts and States Within the Regional Office Jurisdictions

REGION I

Permit Contact, Environmental and Economic Impact Office, U.S. Environmental Protection Agency, John F. Kennedy Building, Boston, Massachusetts 02203, (617) 223—4635, FTS 223—4635.
Connecticut, Maine, Massachusetts, New Hampshire, Rhode Island, and Vermont.

REGION II

Permit Contact, Permits Administration Branch, Room 432, U.S. Environmental Protection Agency, 26 Federal Plaza, New York, New York 10007, (212) 264—9880, FTS 264—9880.
New Jersey, New York, Virgin Islands, and Puerto Rico.

REGION III

Permit Contact *(3 EN 23)*, U.S. Environmental Protection Agency, 6th & Walnut Streets, Philadelphia, Pennsylvania 19106, (215) 597—8816, FTS 597—8816.
Delaware, District of Columbia, Maryland, Pennsylvania, Virginia, and West Virginia.

REGION IV

Permit Contact, Permits Section, U.S. Environmental Protection Agency, 345 Courtland Street, N.E., Atlanta, Georgia 30365, (404) 881—2017, FTS 257—2017.
Alabama, Florida, Georgia, Kentucky, Mississippi, North Carolina, South Carolina, and Tennessee.

REGION V

Permit Contact *(5EP)*, U.S. Environmental Protection Agency, 230 South Dearborn Street, Chicago, Illinois 60604, (312) 353—2105, FTS 353—2105.
Illinois, Indiana, Michigan, Minnesota, Ohio, and Wisconsin.

Table 1 *(continued)*

REGION VI

Permit Contact *(6AEP)*, U.S. Environmental Protection Agency, First International Building, 1201 Elm Street, Dallas, Texas 75270, (214) 767—2765, FTS 729—2765.
  Arkansas, Louisiana, New Mexico, Oklahoma, and Texas.

REGION VII

Permit Contact, Permits Branch, U.S. Environmental Protection Agency, 324 East 11th Street, Kansas City, Missouri 64106, (816) 758—5955, FTS 758—5955.
  Iowa, Kansas, Missouri, and Nebraska.

REGION VIII

Permit Contact *(8E—WE)*, Suite 103, U.S. Environmental Protection Agency, 1860 Lincoln Street, Denver, Colorado 80295, (303) 837—4901, FTS 327—4901.
  Colorado, Montana, North Dakota, South Dakota, Utah, and Wyoming.

REGION IX

Permit Contact, Permits Branch *(E—4)*, U.S. Environmental Protection Agency, 215 Fremont Street, San Francisco, California 94105, (415) 556—3450, FTS 556—3450.
  Arizona, California, Hawaii, Nevada, Guam, American Samoa, and Trust Territories.

REGION X

Permit Contact *(M/S 521)*, U.S. Environmental Protection Agency, 1200 6th Avenue, Seattle, Washington 98101, (206) 442—7176, FTS 399—7176.
  Alaska, Idaho, Oregon, and Washington.

Where to File

The application forms should be mailed to the EPA Regional office whose Region includes the State in which the facility is located *(see Table 1)*.

If the State in which the facility is located administers a Federal permit program under which you need a permit, you should contact the appropriate State agency for the correct forms. Your EPA Regional office *(Table 1)* can tell you to whom to apply and can provide the appropriate address and phone number.

When to File

Because of statutory requirements, the deadlines for filing applications vary according to the type of facility you operate and the type of permit you need. These deadlines are as follows:[1]

Table 2. Filing Dates for Permits

FORM*(permit)*	WHEN TO FILE
2A*(NPDES)*	180 days before your present NPDES permit expires.
2B*(NPDES)*	180 days before your present NPDES permit expires[2], or 180 days prior to startup if you are a new facility.
2C*(NPDES)*	180 days before your present NPDES permit expires[2].
2D*(NPDES)*	180 days prior to startup.
3*(Hazardous Waste)*	Existing facility: Six months following publication of regulations listing hazardous wastes. New facility: 180 days before commencing physical construction.

Table 2 *(continued)*

4*(UIC)*	A reasonable time prior to construction for new wells; as directed by the Director for existing wells.
5*(PSD)*	Prior to commencement of construction.

[1] Please note that some of these forms are not yet available for use and are listed as "Reserved" at the beginning of these instructions. Contact your EPA Regional office for information on current application requirements and forms.

[2] If your present permit expires on or before November 30, 1980, the filing date is the date on which your permit expires. If your permit expires during the period December 1, 1980 — May 31, 1981, the filing date is 90 days before your permit expires.

Federal regulations provide that you may not begin to construct a new source in the NPDES program, a new hazardous waste management facility, a new injection well, or a facility covered by the PSD program before the issuance of a permit under the applicable program. Please note that if you are required to obtain a permit before beginning construction, as described above, you may need to submit your permit application well in advance of an applicable deadline listed in Table 2.

Fees

The U.S. EPA does not require a fee for applying for any permit under the consolidated permit programs. *(However, some States which administer one or more of these programs require fees for the permits which they issue.)*

Availability of Information to Public

Information contained in these application forms will, upon request, be made available to the public for inspection and copying. However, you may request confidential treatment for certain information which you submit on certain supplementary forms. The specific instructions for each supplementary form state what information on the form, if any, may be claimed as confidential and what procedures govern the claim. No information on Forms 1 and 2A through 2D may be claimed as confidential.

Completion of Forms

Unless otherwise specified in instructions to the forms, each item in each form must be answered. To indicate that each item has been considered, enter "NA," for not applicable, if a particular item does not fit the circumstances or characteristics of your facility or activity.

If you have previously submitted information to EPA or to an approved State agency which answers a question, you may either repeat the information in the space provided or attach a copy of the previous submission. Some items in the form require narrative explanation. If more space is necessary to answer a question, attach a separate sheet entitled "Additional Information."

Financial Assistance for Pollution Control

There are a number of direct loans, loan guarantees, and grants available to firms and communities for pollution control expenditures. These are provided by the Small Business Administration, the Economic Development Administration, the Farmers Home Administration, and the Department of Housing and Urban Development. Each EPA Regional office *(Table 1)* has an economic assistance coordinator who can provide you with additional information.

EPA's construction grants program under Title II of the Clean Water Act is an additional source of assistance to publicly owned treatment works. Contact your EPA Regional office for details.

This form must be completed by all applicants.

Completing This Form

Please type or print in the unshaded areas only. Some items have small graduation marks in the fill—in spaces. These marks indicate the number of characters that may be entered into our data system. The marks are spaced at 1/6'' intervals which accommodate elite type *(12 characters per inch)*. If you use another type you may ignore the marks. If you print, place each character between the marks. Abbreviate if necessary to stay within the number of characters allowed for each item. Use one space for breaks between words, but not for punctuation marks unless they are needed to clarify your response.

Item I

Space is provided at the upper right hand corner of Form 1 for insertion of your EPA Identification Number. If you have an existing facility, enter your Identification Number. If you don't know your EPA Identification Number, please contact your EPA Regional office *(Table 1)*, which will provide you with your number. If your facility is new *(not yet constructed)*, leave this item blank.

Item II

Answer each question to determine which supplementary forms you need to fill out. Be sure to check the glossary in Section D of these instructions for the legal definitions of the **bold faced words**. Check Section C of these instructions to determine whether your activity is excluded from permit requirements.

If you answer "no" to every question, then you do not need a permit, and you do not need to complete and return any of these forms.

If you answer "yes" to any question, then you must complete and file the supplementary form by the deadline listed in Table 2 along with this form. *(The applicable form number follows each question and is enclosed in parentheses.)* You need not submit a supplementary form if you already have a permit under the appropriate Federal program, unless your permit is due to expire and you wish to renew your permit.

Questions (I) and (J) of Item II refer to major new or modified sources subject to Prevention of Significant Deterioration *(PSD)* requirements under the Clean Air Act. For the purpose of the PSD program, major sources are defined as: (A) Sources listed in Table 3 which have the potential to emit 100 tons or more per year emissions; and (B) All other sources with the potential to emit 250 tons or more per year. See Section C of these instructions for discussion of exclusions of certain modified sources.

Table 3. 28 Industrial Categories Listed in Section 169(1) of the Clean Air Act of 1977

Fossil fuel—fired steam generators of more than 250 million BTU per hour heat input;
Coal cleaning plants *(with thermal dryers)*;
Kraft pulp mills;
Portland cement plants;
Primary zinc smelters;
Iron and steel mill plants;
Primary aluminum ore reduction plants;
Primary copper smelters;
Municipal incinerators capable of charging more than 250 tons of refuse per day;
Hydrofluoric acid plants;
Nitric acid plants;
Sulfuric acid plants;
Petroleum refineries;
Lime plants;
Phosphate rock processing plants;
Coke oven batteries;
Sulfur recovery plants;
Carbon black plants *(furnace process)*;
Primary lead smelters;
Fuel conversion plants;
Sintering plants;
Secondary metal production plants;
Chemical process plants;
Fossil fuel boilers *(or combination thereof)* totaling more than 250 million BTU per hour heat input;

Table 3 *(continued)*

Petroleum storage and transfer units with a total storage capacity exceeding 300,000 barrels;
Taconite ore processing plants;
Glass fiber processing plants; and
Charcoal production plants.

Item III

Enter the facility's official or legal name. Do not use a colloquial name.

Item IV

Give the name, title, and work telephone number of a person who is thoroughly familiar with the operation of the facility and with the facts reported in this application and who can be contacted by reviewing offices if necessary.

Item V

Give the complete mailing address of the office where correspondence should be sent. This often is not the address used to designate the location of the facility or activity.

Item VI

Give the address or location of the facility identified in Item III of this form. If the facility lacks a street name or route number, give the most accurate alternative geographic information *(e.g., section number or quarter section number from county records or at intersection of Rts. 425 and 22)*.

Item VII

List, in descending order of significance, the four 4—digit standard industrial classification *(SIC)* codes which best describe your facility in terms of the principal products or services you produce or provide. Also, specify each classification in words. These classifications may differ from the SIC codes describing the operation generating the discharge, air emissions, or hazardous wastes.

SIC code numbers are descriptions which may be found in the "Standard Industrial Classification Manual" prepared by the Executive Office of the President, Office of Management and Budget, which is available from the Government Printing Office, Washington, D.C. Use the current edition of the manual. If you have any questions concerning the appropriate SIC code for your facility, contact your EPA Regional office *(see Table 1)*.

Item VIII—A

Give the name, as it is legally referred to, of the person, firm, public organization, or any other entity which operates the facility described in this application. This may or may not be the same name as the facility. The operator of the facility is the legal entity which controls the facility's operation rather than the plant or site manager. Do not use a colloquial name.

Item VIII—B

Indicate whether the entity which operates the facility also owns it by marking the appropriate box.

Item VIII—C

Enter the appropriate letter to indicate the legal status of the operator of the facility. Indicate "public" for a facility solely owned by local government(s) such as a city, town, county, parish, etc.

Items VIII—D — H

Enter the telephone number and address of the operator identified in Item VIII—A.

Item IX

Indicate whether the facility is located on Indian Lands.

Item X

Give the number of each presently effective permit issued to the facility for each program or, if you have previously filed an application but have not yet received a permit, give the number of the application, if any. Fill in the unshaded area only. If you have more than one currently effective permit for your facility under a particular permit program, you may list additional permit numbers on a separate sheet of paper. List any relevant environmental Federal *(e.g., permits under the Ocean Dumping Act, Section 404 of the Clean Water Act or the Surface Mining Control and Reclamation Act)*, State *(e.g., State permits for new air emission sources in nonattainment areas under Part D of the Clean Air Act or State permits under Section 404 of the Clean Water Act)*, or local permits or applications under "other."

Item XI

Provide a topographic map or maps of the area extending at least to one mile beyond the property boundaries of the facility which clearly show the following:

The legal boundaries of the facility;

The location and serial number of each of your existing and proposed intake and discharge structures;

All hazardous waste management facilities;

Each well where you inject fluids underground; and

All springs and surface water bodies in the area, plus all drinking water wells within 1/4 mile of the facility which are identified in the public record or otherwise known to you.

If an intake or discharge structure, hazardous waste disposal site, or injection well associated with the facility is located more than one mile from the plant, include it on the map, if possible. If not, attach additional sheets describing the location of the structure, disposal site, or well, and identify the U.S. Geological Survey *(or other)* map corresponding to the location.

On each map, include the map scale, a meridian arrow showing north, and latitude and longitude at the nearest whole second. On all maps of rivers, show the direction of the current, and in tidal waters, show the directions of the ebb and flow tides. Use a 7-1/2 minute series map published by the U.S. Geological Survey, which may be obtained through the U.S. Geological Survey Offices listed below. If a 7-1/2 minute series map has not been published for your facility site, then you may use a 15 minute series map from the U.S. Geological Survey. If neither a 7-1/2 nor 15 minute series map has been published for your facility site, use a plat map or other appropriate map, including all the requested information; in this case, briefly describe land uses in the map area *(e.g., residential, commercial)*.

You may trace your map from a geological survey chart, or other map meeting the above specifications. If you do, your map should bear a note showing the number or title of the map or chart it was traced from. Include the names of nearby towns, water bodies, and other prominent points. An example of an acceptable location map is shown in Figure 1–1 of these instructions. *(NOTE: Figure 1–1 is provided for purposes of illustration only, and does not represent any actual facility.)*

U.S.G.S. OFFICES	AREA SERVED
Eastern Mapping Center National Cartographic Information Center U.S.G.S. 536 National Center Reston, Va. 22092 Phone No. (703) 860–6336	Ala., Conn., Del., D.C., Fla., Ga., Ind., Ky., Maine, Md., Mass., N.H., N.J., N.Y., N.C., S.C., Ohio, Pa., Puerto Rico, R.I., Tenn., Vt., Va., W. Va., and Virgin Islands.

Item XI *(continued)*

Mid Continent Mapping Center National Cartographic Information Center U.S.G.S. 1400 Independance Road Rolla, Mo. 65401 Phone No. (314) 341–0851	Ark., Ill., Iowa, Kans., La., Mich., Minn., Miss., Mo., N. Dak., Nebr., Okla., S. Dak., and Wis.
Rocky Mountain Mapping Center National Cartographic Infomation Center U.S.G.S. Stop 504, Box 25046 Federal Center Denver, Co. 80225 Phone No. (303) 234–2326	Alaska, Colo., Mont., N. Mex., Tex., Utah, and Wyo.
Western Mapping Center National Cartographic Information Center U.S.G.S. 345 Middlefield Road Menlo Park, Ca. 94025 Phone No. (415) 323–8111	Ariz., Calif., Hawaii, Idaho, Nev., Oreg., Wash., American Samoa, Guam, and Trust Territories

Item XII

Briefly describe the nature of your business *(e.g., products produced or services provided)*.

Item XIII

Federal statues provide for severe penalties for submitting false information on this application form.

18 U.S.C. Section 1001 provides that "Whoever, in any matter within the jurisdiction of any department or agency of the United States knowingly and willfully falsifies, conceals or covers up by any trick, scheme, or device a material fact, or makes or uses any false writing or document knowing same to contain any false, fictitious or fraudulent statement or entry, shall be fined not more than $10,000 or imprisoned not more than five years, or both."

Section 309(c)(2) of the Clean Water Act and Section 113(c)(2) of the Clean Air Act each provide that "Any person who knowingly makes any false statement, representation, or certification in any application, . . . shall upon conviction, be punished by a fine of no more than $10,000 or by imprisonment for not more than six months, or both."

In addition, Section 3008(d)(3) of the Resource Conservation and Recovery Act provides for a fine up to $25,000 per day or imprisonment up to one year, or both, for a first conviction for making a false statement in any application under the Act, and for double these penalties upon subsequent convictions.

FEDERAL REGULATIONS REQUIRE THIS APPLICATION TO BE SIGNED AS FOLLOWS:

A. For a corporation, by a principal executive officer of at least the level of vice president. However, if the only activity in Item II which is marked "yes" is Question G, the officer may authorize a person having responsibility for the overall operations of the well or well field to sign the certification. In that case, the authorization must be written and submitted to the permitting authority.

B. For partnership or sole proprietorship, by a general partner or the proprietor, respectively; or

C. For a municipality, State, Federal, or other public facility, by either a principal executive officer or ranking elected official.

I. National Pollutant Discharge Elimination System Permits Under the Clean Water Act. You are not required to obtain an NPDES permit if your discharge is in one of the following categories, as provided by the Clean Water Act *(CWA)* and by the NPDES regulations *(40 CFR Parts 122—125).* However, under Section 510 of CWA a discharge exempted from the federal NPDES requirements may still be regulated by a State authority; contact your State environmental agency to determine whether you need a State permit.

A. DISCHARGES FROM VESSELS. Discharges of sewage from vessels, effluent from properly functioning marine engines, laundry, shower, and galley sink wastes, and any other discharge incidental to the normal operation of a vessel do not require NPDES permits. However, discharges of rubbish, trash, garbage, or other such materials discharged overboard require permits, and so do other discharges when the vessel is operating in a capacity other than as a means of transportation, such as when the vessel is being used as an energy or mining facility, a storage facility, or a seafood processing facility, or is secured to the bed of the ocean, contiguous zone, or waters of the United States for the purpose of mineral or oil exploration or development.

B. DREDGED OR FILL MATERIAL. Discharges of dredged or fill material into waters of the United States do not need NPDES permits if the dredging or filling is authorized by a permit issued by the U.S. Army Corps of Engineers or an EPA approved State under Section 404 of CWA.

C. DISCHARGES INTO PUBLICLY OWNED TREATMENT WORKS *(POTW).* The introduction of sewage, industrial wastes, or other pollutants into a POTW does not need an NPDES permit. You must comply with all applicable pretreatment standards promulgated under Section 307(b) of CWA, which may be included in the permit issued to the POTW. If you have a plan or an agreement to switch to a POTW in the future, this does not relieve you of the obligation to apply for and receive an NPDES permit until you have stopped discharging pollutants into waters of the United States.

(NOTE: Dischargers into privately owned treatment works do not have to apply for or obtain NPDES permits except as otherwise required by the EPA Regional Administrator. The owner or operator of the treatment works itself, however, must apply for a permit and identify all users in its application. Users so identified will receive public notice of actions taken on the permit for the treatment works.)

D. DISCHARGES FROM AGRICULTURAL AND SILVICULTURAL ACTIVITIES. Most discharges from agricultural and silvicultural activities to waters of the United States do not require NPDES permits. These include runoff from orchards, cultivated crops, pastures, range lands, and forest lands. However, the discharges listed below do require NPDES permits. Definitions of the terms listed below are contained in the Glossary section of these instructions.

1. Discharges from Concentrated Animal Feeding Operations. *(See Glossary for definitions of "animal feeding operations" and "concentrated animal feeding operations." Only the latter require permits.)*

2. Discharges from Concentrated Aquatic Animal Production Facilities. *(See Glossary for size cutoffs.)*

3. Discharges associated with approved Aquaculture Projects.

4. Discharges from Silvicultural Point Sources. *(See Glossary for the definition of "silvicultural point source.")* Nonpoint source silvicultural activities are excluded from NPDES permit requirements. However, some of these activities, such as stream crossings for roads, may involve point source discharges of dredged or fill material which may require a Section 404 permit. See 33 CFR 209.120.

E. DISCHARGES IN COMPLIANCE WITH AN ON—SCENE CO-ORDINATOR'S INSTRUCTIONS.

II. Hazardous Waste Permits Under the Resource Conservation and Recovery Act. You may be excluded from the requirement to obtain a permit under this program if you fall into one of the following categories:

Generators who accumulate their own hazardous waste on—site for less than 90 days as provided in 40 CFR 262.34;

Farmers who dispose of hazardous waste pesticide from their own use as provided in 40 CFR 262.51;

Certain persons treating, storing, or disposing of small quantities of hazardous waste as provided in 40 CFR 261.4 or 261.5; and

Owners and operators of totally enclosed treatment facilities as defined in 40 CFR 260.10.

Check with your Regional office for details. Please note that even if you are excluded from permit requirements, you may be required by Federal regulations to handle your waste in a particular manner.

III. Underground Injection Control Permits Under the Safe Drinking Water Act. You are not required to obtain a permit under this program if you:

Inject into existing wells used to enhance recovery of oil and gas or to store hydrocarbons *(note, however, that these underground injections are regulated by Federal rules)*; or

Inject into or above a stratum which contains, within 1/4 mile of the well bore, an underground source of drinking water *(unless your injection is the type identified in Item II-H, for which you do need a permit)*. However, you must notify EPA of your injection and submit certain required information on forms supplied by the Agency, and your operation may be phased out if you are a generator of hazardous wastes or a hazardous waste management facility which uses wells or septic tanks to dispose of hazardous waste.

IV. Prevention of Significant Deterioration Permits Under the Clean Air Act. The PSD program applies to newly constructed or modified facilities *(both of which are referred to as "new sources")* which increase air emissions. The Clean Air Act Amendments of 1977 exclude small new sources of air emissions from the PSD review program. Any new source in an industrial category listed in Table 3 of these instructions whose potential to emit is less than 100 tons per year is not required to get a PSD permit. In addition, any new source in an industrial category not listed in Table 3 whose potential to emit is less than 250 tons per year is exempted from the PSD requirements.

Modified sources which increase their net emissions *(the difference between the total emission increases and total emission decreases at the source)* less than the significant amount set forth in EPA regulations are also exempt from PSD requirements. Contact your EPA Regional office *(Table 1)* for further information.

NOTE: This Glossary includes terms used in the instructions and in Forms 1, 2B, 2C, and 3. Additional terms will be included in the future when other forms are developed to reflect the requirements of other parts of the Consolidated Permits Program. If you have any questions concerning the meaning of any of these terms, please contact your EPA Regional office *(Table 1)*.

ALIQUOT means a sample of specified volume used to make up a total composite sample.

ANIMAL FEEDING OPERATION means a lot or facility *(other than an aquatic animal production facility)* where the following conditions are met:

A. Animals *(other than aquatic animals)* have been, are, or will be stabled or confined and fed or maintained for a total of 45 days or more in any 12 month period; and

B. Crops, vegetation, forage growth, or post—harvest residues are not sustained in the normal growing season over any portion of the lot or facility.

Two or more animal feeding operations under common ownership are a single animal feeding operation if they adjoin each other or if they use a common area or system for the disposal of wastes.

ANIMAL UNIT means a unit of measurement for any animal feeding operation calculated by adding the following numbers: The number of slaughter and feeder cattle multiplied by 1.0; Plus the number of mature dairy cattle multiplied by 1.4; Plus the number of swine weighing over 25 kilograms *(approximately 55 pounds)* multiplied by 0.4; Plus the number of sheep multiplied by 0.1; Plus the number of horses multiplied by 2.0.

APPLICATION means the EPA standard national forms for applying for a permit, including any additions, revisions, or modifications to the forms; or forms approved by EPA for use in approved States, including any approved modifications or revisions. For RCRA, "application" also means "Application, Part B."

APPLICATION, PART A means that part of the Consolidated Permit Application forms which a RCRA permit applicant must complete to qualify for interim status under Section 3005(e) of RCRA and for consideration for a permit. Part A consists of Form 1 *(General Information)* and Form 3 *(Hazardous Waste Application Form)*.

APPLICATION, PART B means that part of the application which a RCRA permit applicant must complete to be issued a permit. *(NOTE: EPA is not developing a specific form for Part B of the permit application, but an instruction booklet explaining what information must be supplied is available from the EPA Regional office.)*

APPROVED PROGRAM or APPROVED STATE means a State program which has been approved or authorized by EPA under 40 CFR Part 123.

AQUACULTURE PROJECT means a defined managed water area which uses discharges of pollutants into that designated area for the maintenance or production of harvestable freshwater, estuarine, or marine plants or animals. "Designated area" means the portions of the waters of the United States within which the applicant plans to confine the cultivated species, using a method of plan or operation *(including, but not limited to, physical confinement)* which, on the basis of reliable scientific evidence, is expected to ensure the specific individual organisms comprising an aquaculture crop will enjoy increased growth attributable to the discharge of pollutants and be harvested within a defined geographic area.

AQUIFER means a geological formation, group of formations, or part of a formation that is capable of yielding a significant amount of water to a well or spring.

AREA OF REVIEW means the area surrounding an injection well which is described according to the criteria set forth in 40 CFR Section 146.06.

AREA PERMIT means a UIC permit applicable to all or certain wells within a geographic area, rather than to a specified well, under 40 CFR Section 122.37.

ATTAINMENT AREA means, for any air pollutant, an area which has been designated under Section 107 of the Clean Air Act as having ambient air quality levels better than any national primary or secondary ambient air quality standard for that pollutant. Standards have been set for sulfur oxides, particulate matter, nitrogen dioxide, carbon monoxide, ozone, lead, and hydrocarbons. For purposes of the Glossary, "attainment area" also refers to "unclassifiable area," which means, for any pollutants, an area designated under Section 107 as unclassifiable with respect to that pollutant due to insufficient information.

BEST MANAGEMENT PRACTICES *(BMP)* means schedules of activities, prohibitions of practices, maintenance procedures, and other management practices to prevent or reduce the pollution of waters of the United States. BMP's include treatment requirements, operation procedures, and practices to control plant site runoff, spillage or leaks, sludge or waste disposal, or drainage from raw material storage.

BIOLOGICAL MONITORING TEST means any test which includes the use of aquatic algal, invertebrate, or vertebrate species to measure acute or chronic toxicity, and any biological or chemical measure of bioaccumulation.

BYPASS means the intentional diversion of wastes from any any portion of a treatment facility.

CONCENTRATED ANIMAL FEEDING OPERATION means an animal feeding operation which meets the criteria set forth in either (A) or (B) below or which the Director designates as such on a case—by—case basis:

A. More than the numbers of animals specified in any of the following categories are confined:

1. 1,000 slaughter or feeder cattle,

2. 700 mature dairy cattle *(whether milked or dry cows)*,

3. 2,500 swine each weighing over 25 kilograms *(approximately 55 pounds)*,

4. 500 horses,

5. 10,000 sheep or lambs,

6. 55,000 turkeys,

7. 100,000 laying hens or broilers *(if the facility has a continuous overflow watering)*,

8. 30,000 laying hens or broilers *(if the facility has a liquid manure handling system)*,

9. 5,000 ducks, or

10. 1,000 animal units; or

B. More than the following numbers and types of animals are confined:

1. 300 slaughter or feeder cattle,

2. 200 mature dairy cattle *(whether milked or dry cows)*,

3. 750 swine each weighing over 25 kilograms *(approximately 55 pounds)*,

4. 150 horses,

CONCENTRATED ANIMAL FEEDING OPERATION *(continued)*

   5. 3,000 sheep or lambs,

   6. 16,500 turkeys,

   7. 30,000 laying hens or broilers *(if the facility has continuous overflow watering)*,

   8. 9,000 laying hens or broilers *(if the facility has a liquid manure handling system)*,

   9. 1,500 ducks, or

   10. 300 animal units; AND

Either one of the following conditions are met: Pollutants are discharged into waters of the United States through a manmade ditch, flushing system or other similar manmade device *("manmade" means constructed by man and used for the purpose of transporting wastes)*; or Pollutants are discharged directly into waters of the Unites States which originate outside of and pass over, across, or through the facility or otherwise come into direct contact with the animals confined in the operation.

Provided, however, that no animal feeding operation is a concentrated animal feeding operation as defined above if such animal feeding operation discharges only in the event of a 25 year, 24 hour storm event.

CONCENTRATED AQUATIC ANIMAL PRODUCTION FACILITY means a hatchery, fish farm, or other facility which contains, grows or holds aquatic animals in either of the following categories, or which the Director designates as such on a case—by—case basis:

A. Cold water fish species or other cold water aquatic animals including, but not limited to, the Salmonidae family of fish *(e.g., trout and salmon)* in ponds, raceways or other similar structures which discharge at least 30 days per year but does not include:

   1. Facilities which produce less than 9,090 harvest weight kilograms *(approximately 20,000 pounds)* of aquatic animals per year; and

   2. Facilities which feed less than 2,272 kilograms *(approximately 5,000 pounds)* of food during the calendar month of maximum feeding.

B. Warm water fish species or other warm water aquatic animals including, but not limited to, the Ameiuridae, Cetrarchidae, and Cyprinidae families of fish *(e.g., respectively, catfish, sunfish, and minnows)* in ponds, raceways, or other similar structures which discharge at least 30 days per year, but does not include:

   1. Closed ponds which discharge only during periods of excess runoff; or

   2. Facilities which produce less than 45,454 harvest weight kilograms *(approximately 100,000 pounds)* of aquatic animals per year.

CONTACT COOLING WATER means water used to reduce temperature which comes into contact with a raw material, intermediate product, waste product other than heat, or finished product.

CONTAINER means any portable device in which a material is stored, transported, treated, disposed of, or otherwise handled.

CONTIGUOUS ZONE means the entire zone established by the United States under article 24 of the convention of the Territorial Sea and the Contiguous Zone.

CWA means the Clean Water Act *(formerly referred to the Federal Water Pollution Control Act)* Pub. L. 92–500, as amended by Pub. L. 95–217 and Pub. L. 95–576, 33 U.S.C. 1251 *et seq.*

DIKE means any embankment or ridge of either natural or manmade materials used to prevent the movement of liquids, sludges, solids, or other materials.

DIRECT DISCHARGE means the discharge of a pollutant as defined below.

DIRECTOR means the EPA Regional Administrator or the State Director as the context requires.

DISCHARGE *(OF A POLLUTANT)* means:

A. Any addition of any pollutant or combination of pollutants to waters of the United States from any point source; or

B. Any addition of any pollutant or combination of pollutants to the waters of the contiguous zone or the ocean from any point source other than a vessel or other floating craft which is being used as a means of transportation.

This definition includes discharges into waters of the United States from: Surface runoff which is collected or channelled by man; Discharges through pipes, sewers, or other conveyances owned by a State, municipality, or other person which do not lead to POTW's; and Discharges through pipes, sewers, or other conveyances, leading into privately owned treatment works. This term does not include an addition of pollutants by any indirect discharger.

DISPOSAL *(in the RCRA program)* means the discharge, deposit, injection, dumping, spilling, leaking, or placing of any hazardous waste into or on any land or water so that the hazardous waste or any constituent of it may enter the environment or be emitted into the air or discharged into any waters, including ground water.

DISPOSAL FACILITY means a facility or part of a facility at which hazardous waste is intentionally placed into or on land or water, and at which hazardous waste will remain after closure.

EFFLUENT LIMITATION means any restriction imposed by the Director on quantities, discharge rates, and concentrations of pollutants which are discharged from point sources into waters of the United States, the waters of the contiguous zone, or the ocean.

EFFLUENT LIMITATION GUIDELINE means a regulation published by the Administrator under Section 304(b) of the Clean Water Act to adopt or revise effluent limitations.

ENVIRONMENTAL PROTECTION AGENCY *(EPA)* means the United States Environmental Protection Agency.

EPA IDENTIFICATION NUMBER means the number assigned by EPA to each generator, transporter, and facility.

EXEMPTED AQUIFER means an aquifer or its portion that meets the criteria in the definition of USDW, but which has been exempted according to the procedures in 40 CFR Section 122.35(b).

EXISTING HWM FACILITY means a Hazardous Waste Management facility which was in operation, or for which construction had commenced, on or before October 21, 1976. Construction had commenced if (A) the owner or operator had obtained all necessary Federal, State, and local preconstruction approvals or permits, and either (B1) a continuous on—site, physical construction program had begun, or (B2) the owner or operator had entered into contractual obligations, which could not be cancelled or modified without substantial loss, for construction of the facility to be completed within a reasonable time.

(NOTE: This definition reflects the literal language of the statute. However, EPA believes that amendments to RCRA now in conference will shortly be enacted and will change the date for determining when a facility is an "existing facility" to one no earlier than May of 1980; indications are the conferees are considering October 30, 1980. Accordingly, EPA encourages every owner or operator of a facility which was built or under construction as of the promulgation date of the RCRA program regulations to file Part A of its permit application so that it can be quickly processed for interim status when the change in the law takes effect. When those amendments are enacted, EPA will amend this definition.)

EXISTING SOURCE or EXISTING DISCHARGER *(in the NPDES program)* means any source which is not a new source or a new discharger.

EXISTING INJECTION WELL means an injection well other than a new injection well.

FACILITY means any HWM facility, UIC underground injection well, NPDES point source, PSD stationary source, or any other facility or activity *(including land or appurtenances thereto)* that is subject to regulation under the RCRA, UIC, NPDES, or PSD programs.

FLUID means material or substance which flows or moves whether in a semisolid, liquid, sludge, gas, or any other form or state.

GENERATOR means any person by site, whose act or process produces hazardous waste identified or listed in 40 CFR Part 261.

GROUNDWATER means water below the land surface in a zone of saturation.

HAZARDOUS SUBSTANCE means any of the substances designated under 40 CFR Part 116 pursuant to Section 311 of CWA. *(NOTE: These substances are listed in Table 2c—4 of the instructions to Form 2C.)*

HAZARDOUS WASTE means a hazardous waste as defined in 40 CFR Section 261.3 published May 19, 1980.

HAZARDOUS WASTE MANAGEMENT FACILITY *(HWM facility)* means all contiguous land, structures, appurtenances, and improvements on the land, used for treating, storing, or disposing of hazardous wastes. A facility may consist of several treatment, storage, or disposal operational units *(for example, one or more landfills, surface impoundments, or combinations of them)*.

IN OPERATION means a facility which is treating, storing, or disposing of hazardous waste.

INCINERATOR *(in the RCRA program)* means an enclosed device using controlled flame combustion, the primary purpose of which is to thermally break down hazardous waste. Examples of incinerators are rotary kiln, fluidized bed, and liquid injection incinerators.

INDIRECT DISCHARGER means a nondomestic discharger introducing pollutants to a publicly owned treatment works.

INJECTION WELL means a well into which fluids are being injected.

INTERIM AUTHORIZATION means approval by EPA of a State hazardous waste program which has met the requirements of Section 3006(c) of RCRA and applicable requirements of 40 CFR Part 123, Subparts A, B, and F.

LANDFILL means a disposal facility or part of a facility where hazardous waste is placed in or on land and which is not a land treatment facility, a surface impoundment, or an injection well.

LAND TREATMENT FACILITY *(in the RCRA program)* means a facility or part of a facility at which hazardous waste is applied onto or incorporated into the soil surface; such facilities are disposal facilities if the waste will remain after closure.

LISTED STATE means a State listed by the Administrator under Section 1422 of SDWA as needing a State UIC program.

MGD means millions of gallons per day.

MUNICIPALITY means a city, village, town, borough, county, parish, district, association, or other public body created by or under State law and having jurisdiction over disposal of sewage, industrial wastes, or other wastes, or an Indian tribe or an authorized Indian tribal organization, or a designated and approved management agency under Section 208 of CWA.

NATIONAL POLLUTANT DISCHARGE ELIMINATION SYSTEM *(NPDES)* means the national program for issuing modifying, revoking and reissuing, terminating, monitoring, and enforcing permits and imposing and enforcing pretreatment requirements, under Sections 307, 318, 402, and 405 of CWA. The term includes an approved program.

NEW DISCHARGER means any building, structure, facility, or installation: (A) From which there is or may be a new or additional discharge of pollutants at a site at which on October 18, 1972, it had never discharged pollutants; (B) Which has never received a finally effective NPDES permit for discharges at that site; and (C) Which is not a "new source." This definition includes an indirect discharger which commences discharging into waters of the United States. It also includes any existing mobile point source, such as an offshore oil drilling rig, seafood processing vessel, or aggregate plant that begins discharging at a location for which it does not have an existing permit.

NEW HWM FACILITY means a Hazardous Waste Management facility which began operation or for which construction commenced after October 21, 1976.

NEW INJECTION WELL means a well which begins injection after a UIC program for the State in which the well is located is approved.

NEW SOURCE *(in the NPDES program)* means any building, structure, facility, or installation from which there is or may be a discharge of pollutants, the construction of which commenced:

A. After promulgation of standards of performance under Section 306 of CWA which are applicable to such source; or

B. After proposal of standards of performance in accordance with Section 306 of CWA which are applicable to such source, but only if the standards are promulgated in accordance with Section 306 within 120 days of their proposal.

NON—CONTACT COOLING WATER means water used to reduce temperature which does not come into direct contact with any raw material, intermediate product, waste product *(other than heat)*, or finished product.

OFF—SITE means any site which is not "on—site."

ON—SITE means on the same or geographically contiguous property which may be divided by public or private right*(s)*—of—way, provided the entrance and exit between the properties is at a cross—roads intersection, and access is by crossing as opposed to going along, the right*(s)*—of—way. Non—contiguous properties owned by the same person, but connected by a right—of—way which the person controls and to which the public does not have access, is also considered on—site property.

OPEN BURNING means the combustion of any material without the following characteristics:

A. Control of combustion air to maintain adequate temperature for efficient combustion;

B. Containment of the combustion—reaction in an enclosed device to provide sufficient residence time and mixing for complete combustion; and

C. Control of emission of the gaseous combustion products.

(See also "incinerator" and "thermal treatment").

OPERATOR means the person responsible for the overall operation of a facility.

OUTFALL means a point source.

OWNER means the person who owns a facility or part of a facility.

PERMIT means an authorization, license, or equivalent control document issued by EPA or an approved State to implement the requirements of 40 CFR Parts 122, 123, and 124.

PHYSICAL CONSTRUCTION *(in the RCRA program)* means excavation, movement of earth, erection of forms or structures, or similar activity to prepare a HWM facility to accept hazardous waste.

PILE means any noncontainerized accumulation of solid, nonflowing hazardous waste that is used for treatment or storage.

POINT SOURCE means any discernible, confined, and discrete conveyance, including but not limited to any pipe, ditch, channel, tunnel, conduit, well, discrete fissure, container, rolling stock, concentrated animal feeding operation, vessel or other floating craft from which pollutants are or may be discharged. This term does not include return flows from irrigated agriculture.

POLLUTANT means dredged spoil, solid waste, incinerator residue, filter backwash, sewage, garbage, sewage sludge, munitions, chemical waste, biological materials, radioactive materials *(except those regulated under the Atomic Energy Act of 1954, as amended [42 U.S.C. Section 2011 et seq.])*, heat, wrecked or discarded equipment, rocks, sand, cellar dirt and industrial, municipal, and agriculture waste discharged into water. It does not mean:

A. Sewage from vessels; or

B. Water, gas, or other material which is injected into a well to facilitate production of oil or gas, or water derived in association with oil and gas production and disposed of in a well, if the well used either to facilitate production or for disposal purposes is approved by authority of the State in which the well is located, and if the State determines that the injection or disposal will not result in the degradation of ground or surface water resources.

(NOTE: Radioactive materials covered by the Atomic Energy Act are those encompassed in its definition of source, byproduct, or special nuclear materials. Examples of materials not covered include radium and accelerator produced isotopes. See Train v. Colorado Public Interest Research Group, Inc., 426 U.S. 1 [1976].)

PREVENTION OF SIGNIFICANT DETERIORATION *(PSD)* means the national permitting program under 40 CFR 52.21 to prevent emissions of certain pollutants regulated under the Clean Air Act from significantly deteriorating air quality in attainment areas.

PRIMARY INDUSTRY CATEGORY means any industry category listed in the NRDC Settlement Agreement *(Natural Resources Defense Council v. Train, 8 ERC 2120 [D.D.C. 1976], modified 12 ERC 1833 [D.D.C. 1979])*.

PRIVATELY OWNED TREATMENT WORKS means any device or system which is: (A) Used to treat wastes from any facility whose operator is not the operator of the treatment works; and (B) Not a POTW.

PROCESS WASTEWATER means any water which, during manufacturing or processing, comes into direct contact with or results from the production or use of any raw material, intermediate product, finished product, byproduct, or waste product.

PUBLICLY OWNED TREATMENT WORKS or POTW means any device or system used in the treatment *(including recycling and reclamation)* of municipal sewage or industrial wastes of a liquid nature which is owned by a State or municipality. This definition includes any sewers, pipes, or other conveyances only if they convey wastewater to a POTW providing treatment.

RENT means use of another's property in return for regular payment.

RCRA means the Solid Waste Disposal Act as amended by the Resource Conservation and Recovery Act of 1976 *(Pub. L. 94–580, as amended by Pub. L. 95–609, 42 U.S.C. Section 6901 et seq.)*.

ROCK CRUSHING AND GRAVEL WASHING FACILITIES are facilities which process crushed and broken stone, gravel, and riprap *(see 40 CFR Part 436, Subpart B, and the effluent limitations guidelines for these facilities)*.

SDWA means the Safe Drinking Water Act *(Pub. L. 95–523, as amended by Pub. L. 95–1900, 42 U.S.C. Section 300[f] et seq.)*.

SECONDARY INDUSTRY CATEGORY means any industry category which is not a primary industry category.

SEWAGE FROM VESSELS means human body wastes and the wastes from tiolets and other receptacles intended to receive or retain body wastes that are discharged from vessels and regulated under Section 312 of CWA, except that with respect to commercial vessels on the Great Lakes this term includes graywater. For the purposes of this definition, "graywater" means galley, bath, and shower water.

SEWAGE SLUDGE means the solids, residues, and precipitate separated from or created in sewage by the unit processes of a POTW. "Sewage" as used in this definition means any wastes, including wastes from humans, households, commercial establishments, industries, and storm water runoff, that are discharged to or otherwise enter a publicly owned treatment works.

SILVICULTURAL POINT SOURCE means any discernable, confined, and discrete conveyance related to rock crushing, gravel washing, log sorting, or log storage facilities which are operated in connection with silvicultural activities and from which pollutants are discharged into waters of the United States. This term does not include nonpoint source silvicultural activities such as nursery operations, site preparation, reforestation and subsequent cultural treatment, thinning, prescribed burning, pest and fire control, harvesting operations, surface drainage, or road construction and maintenance from which there is natural runoff. However, some of these activities *(such as stream crossing for roads)* may involve point source discharges of dredged or fill material which may require a CWA Section 404 permit. "Log sorting and log storage facilities" are facilities whose discharges result from the holding of unprocessed wood, e.g., logs or roundwood with bark or after removal of bark in self–contained bodies of water *(mill ponds or log ponds)* or stored on land where water is applied intentionally on the logs *(wet decking)*. (See 40 CFR Part 429, Subpart J, and the effluent limitations guidelines for these facilities.)

STATE means any of the 50 States, the District of Columbia, Guam, the Commonwealth of Puerto Rico, the Virgin Islands, American Samoa, the Trust Territory of the Pacific Islands *(except in the case of RCRA)*, and the Commonwealth of the Northern Mariana Islands *(except in the case of CWA)*.

STATIONARY SOURCE *(in the PSD program)* means any building, structure, facility, or installation which emits or may emit any air pollutant regulated under the Clean Air Act. "Building, structure, facility, or installation" means any grouping of pollutant—emitting activities which are located on one or more contiguous or adjacent properties and which are owned or operated by the same person *(or by persons under common control)*.

STORAGE *(in the RCRA program)* means the holding of hazardous waste for a temporary period at the end of which the hazardous waste is treated, disposed, or stored elsewhere.

STORM WATER RUNOFF means water discharged as a result of rain, snow, or other precipitation.

SURFACE IMPOUNDMENT or IMPOUNDMENT means a facility or part of a facility which is a natural topographic depression, manmade excavation, or diked area formed primarily of earthen materials *(although it may be lined with manmade materials)*, which is designed to hold an accumulation of liquid wastes or wastes containing free liquids, and which is not an injection well. Examples of surface impoundments are holding, storage, settling, and aeration pits, ponds, and lagoons.

TANK *(in the RCRA program)* means a stationary device, designed to contain an accumulation of hazardous waste which is constructed primarily of non–earthen materials *(e.g., wood, concrete, steel, plastic)* which provide structural support.

THERMAL TREATMENT *(in the RCRA program)* means the treatment of hazardous waste in a device which uses elevated temperature as the primary means to change the chemical, physical, or biological character or composition of the hazardous waste. Examples of thermal treatment processes are incineration, molten salt, pyrolysis, calcination, wet air oxidation, and microwave discharge. *(See also "incinerator" and "open burning").*

TOTALLY ENCLOSED TREATMENT FACILITY *(in the RCRA program)* means a facility for the treatment of hazardous waste which is directly connected to an industrial production process and which is constructed and operated in a manner which prevents the release of any hazardous waste or any constituent thereof into the environment during treatment. An example is a pipe in which waste acid is neutralized.

TOXIC POLLUTANT means any pollutant listed as toxic under Section 307(a)(1) of CWA.

TRANSPORTER *(in the RCRA program)* means a person engaged in the off—site transportation of hazardous waste by air, rail, highway, or water.

TREATMENT *(in the RCRA program)* means any method, technique, or process, including neutralization, designed to change the physical, chemical, or biological character or composition of any hazardous waste so as to neutralize such waste, or so as to recover energy or material resources from the waste, or so as to render such waste non—hazardous, or less hazardous; safer to transport, store, or dispose of; or amenable for recovery, amenable for storage, or reduced in volume.

UNDERGROUND INJECTION means well injection.

UNDERGROUND SOURCE OF DRINKING WATER or USDW means an aquifer or its portion which is not an exempted aquifer and:

A. Which supplies drinking water for human consumption; or

B. In which the ground water contains fewer than 10,000 mg/l total dissolved solids.

UPSET means an exceptional incident in which there is unintentional and temporary noncompliance with technology—based permit effluent limitations because of factors beyond the reasonable control of the permittee. An upset does not include noncompliance to the extent caused by operational error, improperly designed treatment facilities, inadequate treatment facilities, lack of preventive maintenance, or careless or improper operation.

WATERS OF THE UNITED STATES means:

A. All waters which are currently used, were used in the past, or may be susceptible to use in interstate or foreign commerce, including all waters which are subject to the ebb and flow of the tide;

B. All interstate waters, including interstate wetlands;

C. All other waters such as intrastate lakes, rivers, streams *(including intermittent streams)*, mudflats, sandflats, wetlands, sloughs, prairie potholes, wet meadows, playa lakes, and natural ponds, the use, degradation, or destruction of which would or could affect interstate or foreign commerce including any such waters:

1. Which are or could be used by interstate or foreign travelers for recreational or other purposes,

2. From which fish or shellfish are or could be taken and sold in interstate or foreign commerce,

3. Which are used or could be used for industrial purposes by industries in interstate commerce;

D. All impoundments of waters otherwise defined as waters of the United States under this definition;

E. Tributaries of waters identified in paragraphs (A) — (D) above;

F. The territorial sea; and

G. Wetlands adjacent to waters *(other than waters that are themselves wetlands)* identified in paragraphs (A) — (F) of this definition.

Waste treatment systems, including treatment ponds or lagoons designed to meet requirement of CWA *(other than cooling ponds as defined in 40 CFR Section 423.11(m) which also meet the criteria of this definition)* are not waters of the United States. This exclusion applies only to manmade bodies of water which neither were originally created in waters of the United States *(such as a disposal area in wetlands)* nor resulted from the impoundments of waters of the United States.

WELL INJECTION or UNDERGROUND INJECTION means the subsurface emplacement of fluids through a bored, drilled, or driven well; or through a dug well, where the depth of the dug well is greater than the largest surface dimension.

WETLANDS means those areas that are inundated or saturated by surface or groundwater at a frequency and duration sufficient to support, and that under normal circumstances do support, a prevalence of vegetation typically adapted for life in saturated soil conditions. Wetlands generally include swamps, marshes, bogs, and similar areas.

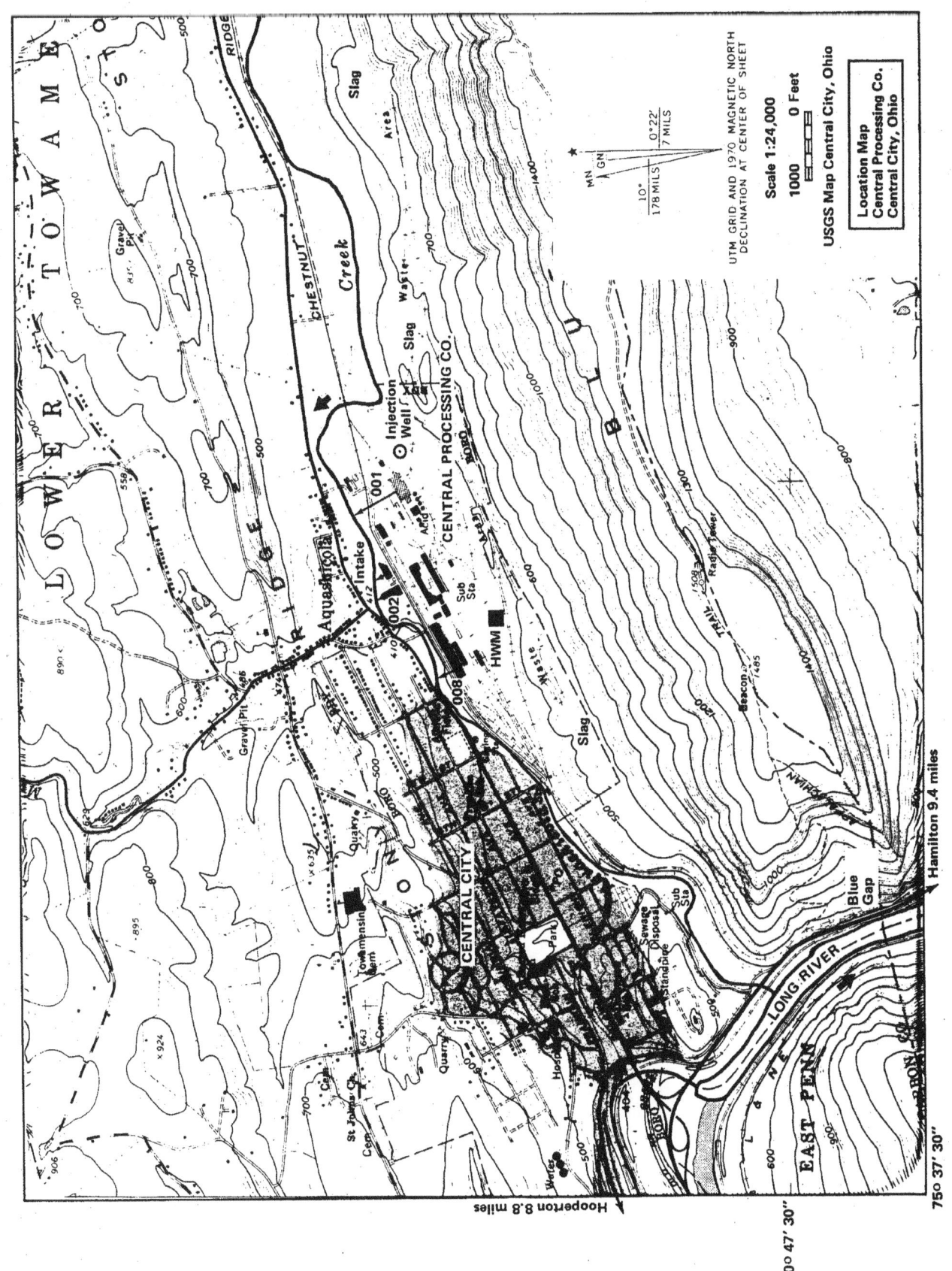

FIGURE 1-1

Appendix K2

NPDES Permit Applications: Form 2B

APPENDIX - FORM 2B

Form Approved
OMB No, 2040-0250
Approval expires 12-15-05

EPA I.D. NUMBER *(copy from Item 1 of Form 1)*

FORM **2B** NPDES	**EPA**	U.S. ENVIRONMENTAL PROTECTION AGENCY APPLICATIONS FOR PERMIT TO DISCHARGE WASTEWATER CONCENTRATED ANIMAL FEEDING OPERATIONS AND AQUATIC ANIMAL PRODUCTION FACILITIES

I. GENERAL INFORMATION   Applying for: Individual Permit ☐   Coverage Under General Permit ☐

A. TYPE OF BUSINESS	B. CONTACT INFORMATION	C. FACILITY OPERATION STATUS
☐ 1. Concentrated Animal Feeding Operation (complete items B, C, D, and Section II) ☐ 2. Concentrated Aquatic Animal Production Facility (complete items B, C, and section III)	Owner/or Operator Name: _____ Telephone: (_____) _____ Address: _____ Facsimile: (_____) _____ City: _____ State: __ Zip Code: _____	☐ 1. Existing Facility ☐ 2. Proposed Facility

A. FACILITY INFORMATION

Name: _____ Telephone: (____) _____
Address: _____ Facsimile: (____) _____
City: _____ State: _____ Zip Code: _____
County: _____ Latitude: _____ Longitude: _____

If contract operation: Name of Integrator: _____
Address of Integrator: _____

II. CONCENTRATED ANIMAL FEEDING OPERATION CHARACTERISTICS

A. TYPE AND NUMBER OF ANIMALS			B. Manure, Litter and/or Wastewater Production and Use
	2. ANIMALS		a) How much manure, litter and wastewater is generated annually by the facility? _____ tons _____ gallons
1. TYPE	NO. IN OPEN CONFINEMENT	NO. HOUSED UNDER ROOF	b) If land applied how many acres of land under the control of the applicant are available for applying the CAFOs manure/litter/wastewater? _____ acres
☐ Mature Dairy Cows			c) How many tons of manure or litter, or gallons of waste-water produced by the CAFO will be transferred annually to other persons? tons/gallons *(circle one)*
☐ Dairy Heifers			
☐ Veal Calves			
☐ Cattle (not dairy or veal)			
☐ Swine (55 lb. or over)			
☐ Swine (under 55 lb.)			
☐ Horses			
☐ Sheep or Lambs			

EPA Form 3510-2B (12-02)

Form Approved
OMB No, 2040-0250
Approval expires 12-15-05

❑ Turkeys			
❑ Chickens (Broilers)			
❑ Chickens (Layers)			
❑ Ducks			
❑ Other Specify _____			
3. TOTAL ANIMALS			

C. ❑ TOPOGRAPHIC MAP

D. TYPE OF CONTAINMENT, STORAGE AND CAPACITY

1. Type of Containment	Total Capacity (in gallons)	
❑ Lagoon		
❑ Holding Pond		
❑ Evaporation Pond		
❑ Other: Specify _____		

2. Report the total number of acres contributing drainage: _____ acres

3. Type of Storage	Total Number. of Days	Total Capacity (gallons/tons)	
❑ Anaerobic Lagoon			
❑ Storage Lagoon			
❑ Evaporation Pond			
❑ Aboveground Storage Tanks			
❑ Belowground Storage Tanks			
❑ Roofed Storage Shed			
❑ Concrete Pad			
❑ Impervious Soil Pad			
❑ Other: Specify _____			

EPA Form 3510-2B (12-02)

Form Approved
OMB No, 2040-0250
Approval expires 12-15-05

E. NUTRIENT MANAGEMENT PLAN

   A. Has a nutrient management plan been developed? ☐ Yes ☐ No

   B. Is a nutrient management plan being implemented for the facility? ☐ Yes ☐ No

   C. If no, when will the nutrient management plan be developed? Date: _____

   D. The date of the last review or revision of the nutrient management plan. Date: _____

   E. If not land applying, describe alternative use(s) of manure, litter and or wastewater:

F. LAND APPLICATION BEST MANAGEMENT PRACTICES
Please check any of the following best management practices that are being implemented at the facility to control runoff and protect water quality:

   ☐ Buffers ☐ Setbacks ☐ Conservation tillage ☐ Constructed wetlands ☐ Infiltration field ☐ Grass filter
   ☐ Terrace

III. CONCENTRATED AQUATIC ANIMAL PRODUCTION FACILITY CHARACTERISTICS

A. For each outfall give the maximum daily flow, maximum 30- day flow, and the long-term average flow.				B. Indicate the total number of ponds, raceways, and similar structures in your facility.		
1. Outfall No.	2. Flow (*gallons per day*)			1. Ponds	2. Raceways	3. Other
	a. Maximum. Daily	b. Maximum 30 Day	c. Long Term Average	C. Provide the name of the receiving water and the source of water used by your facility.		
				1. Receiving Water	2. Water Source	

D. List the species of fish or aquatic animals held and fed at your facility. For each species, give the total weight produced by your facility per year in pounds of harvestable weight, and also give the maximum weight present at any one time.

1. Cold Water Species				2. Warm Water Species			
a. Species	b. Harvestable Weight (*pounds*)			a. Species	b. Harvestable Weight (*pounds*)		
		(1) Total Yearly	(2) Maximum			(1) Total Yearly	(2) Maximum

E. Report the total pounds of food during the calendar month of maximum feeding.	1. Month	2. Pounds of Food

Form Approved
OMB No, 2040-0250
Approval expires 12-15-05

IV. CERTIFICATION	
I certify under penalty of law that I have personally examined and am familiar with the information submitted in this application and all attachments and that, based on my inquiry of those individuals immediately responsible for obtaining the information, I believe that the information is true accurate and complete. I am aware that there are significant penalties for submitting false information, including the possibility of fine and imprisonment.	
A. Name and Official Title *(print or type)*	B. Phone No. ( )
C. Signature	D. Date Signed

EPA Form 3510-2B (12-02)

Form Approved
OMB No, 2040-0250
Approval expires 12-15-05

INSTRUCTIONS

GENERAL

This form must be completed by all applicants who check "yes" to Item II-B in Form 1. Not all animal feeding operations or fish farms are required to obtain NPDES permits. Exclusions are based on size. See the description of these statutory and regulatory exclusions in the General Instructions that accompany Form 1.

For aquatic animal production facilities, the size cutoffs are based on whether the species are warm water or cold water, on the production weight per year in harvestable pounds, and on the amount of feeding in pounds of food (*for cold water species*). Also, facilities which discharge less than 30 days per year, or only during periods of excess runoff (*for warm water fish*) are not required to have a permit.

Refer to the Form 1 instructions to determine where to file this form.

Item I-A

See the note above and the General Instructions which accompany Form 1 to be sure that your facility is a "concentrated animal feeding operation" (CAFO).

Item I-B

Use this space to give owner/operator contact information.

Item I-C

Check "proposed" if your facility is not now in operation or is expanding to meet the definition of a CAFO in accordance with the information found in the General Instructions that accompany Form 1.

Item I-D

Use this space to give a complete legal description of your facility's location including name, address, and latitude/longitude. Also, the if a contract grower, the name and address of the integrator.

Item II

Supply all information in item II if you checked (1) in item I-A.

Item II-A

Give the maximum number of each type of animal in open confinement or housed under roof (either partially or totally) which are held at your facility for a total of 45 days or more in any 12 month period. Provide the total number of animals confined at the facility.

Item II-B

Provide the total amount of manure, litter and wastewater generated annually by the facility. Identify if manure, litter and wastewater generated by the facility is to be land applied and the number of acres, under the control of the CAFO operator, suitable for land application. If the answer to question 3 is yes, provide the estimated annual quantity of manure, litter and wastewater that the applicant plans to transfer off-site.

Item II-C

Check this box if you have submitted a topographic map of the geographic area in which the CAFO is located showing the specific location of the production area.

Item II-D

1. Provide information on the type of containment and the capacity of the containment structure (s).
2. The number of acres that are drained and collected in the containment structure (s).
3. Identify the type of storage for the manure, litter and/or wastewater. Give the capacity of this storage in days and gallons or tons.

Item II-E

Provide information concerning the status of the development and implementation of a nutrient management plan for the facility. In those cases where the nutrient management plan has not been completed, provide an estimated date of development and implementation. If not land applying, describe the alternative uses of the manure, litter and wastewater (e.g., composting, pelletizing, energy generation, etc.).

Item II-F

Check any of the identified conservation practices that are being implemented at the facility to control runoff and protect water quality.

Item III

Supply all information in Item III if you checked (2) in Item I-A.

Item III-A

Outfalls should be numbered to correspond with the map submitted in Item XI of Form 1. Values given for flow should be representative of your normal operation. The maximum daily flow is the maximum measured flow occurring over a calendar day. The maximum 30-day flow is the average of measured daily flow over the calendar month of highest flow. The long-term average flow is the average of measure daily flows over a calendar year.

Item III-B

Give the total number of discrete ponds or raceways in your facility. Under "other," give a descriptive name of any structure which is not a pond or a raceway but which results in discharge to waters of the United States.

Item III-C

Use names for receiving water and source of water which correspond to the map submitted in Item XI of Form 1.

Item III-D

The names of fish species should be proper, common, or scientific names as given in special Publication No. 6 of the American Fisheries Society. "A List of Common and Scientific Names of Fishes from the United States and Canada." The values given for total weight produced by your facility per year and the maximum weight present at any one time should be representative of your normal operation.

Item III-E

The value given for maximum monthly pounds of food should be representative of your normal operation.

Item IV

The Clean Water Act provides for severe penalties for submitting false information on this application form.

Section 309(C)(2) of the Clean Water Act provides that "Any person who knowingly makes any false statement, representation, or certification in any application...shall upon conviction, be punished by a fine of no more than $10,000 or by imprisonment for not more than six months, or both."

Federal regulations require the certification to be signed as follows:
A. For corporation, by a principal executive officer of at least the level of vice president.
B. For a partnership or sole proprietorship, by a general partner or the proprietor, respectively; or
C. For a municipality, State, Federal, or other public facility, by either a principal executive officer or ranking elected official.

EPA Form 3510-2B (12-02)

Appendix L

Applicability Matrix

APPLICABILITY MATRIX

System Type	Species Type (Water)	Discharge > than 30 Days Per Year?	Annual Production of Aquatic Animals	Maximum Feeding is > than 5,000 lb (2,272 kg)?	NPDES Applies?[1]	NPDES / ELGs Applies?
Flow-through or Recirculating	Cold	Yes	≥ 100,000 lb	N/A	X	X
			≥ 20,000 lb (9,090 kg)	Yes	X	
				No		
			< 20,000 lb (9,090 kg)			
		No				
	Warm	Yes	≥ 100,000 lb	N/A	X	X
			≥ 100,000 lb (45,454 kg)	Yes	X	X
				No		
			< 100,000 lb (45,454 kg)			
		No				
Net Pens	Cold	Yes	≥ 100,000 lb	N/A	X	X
			≥ 20,000 lb (9,090 kg)	Yes	X	
				No		
			< 20,000 lb (9,090 kg)			
		No				
	Warm	Yes	≥ 100,000 lb	N/A	X	X
			≥ 100,000 lb (45,454 kg)	Yes	X	X
				No		
			< 100,000 lb (45,454 kg)			
		No				
Ponds	Cold	Yes	≥ 20,000 lb (9,090 kg)	Yes	X	
				No		
			< 20,000 lb (9,090 kg)			
		No				
	Warm	Yes	≥ 20,000 lb (9,090 kg)	Yes	X	
				No		
			< 20,000 lb (9,090 kg)			
		No				
Alligator ponds, molluscan shellfish, lobster cages and pounds, crawfish, indirect dischargers, or Alaskan flow-through[2]					See footnote[2]	

[1] The Director may designate a facility as a CAAP facility on a case-by-case basis, even if the facility does not meet the discharge, annual production, and feed requirements of the NPDES regulations.

[2] These types of systems are exempt from the CAAP ELGs. They may be regulated by the NPDES regulations if they meet the discharge, annual production, and feed requirements of the NPDES regulations, or if the Director designates them (on a case-by-case basis) as CAAP facilities or other types of facilities requiring an NPDES permit.

Appendix M1

Example Written Report
Participating in an INAD Study

EXAMPLE WRITTEN REPORT FOR AGREEING
TO PARTICIPATE IN AN INAD STUDY

(Submit a written report to your permitting authority within
7 days of agreeing or signing up to participate in an INAD study)

Facility Name:_____   NPDES Permit Number:_____

Name of person submitting this report:_____

Date this written report was submitted to the permitting authority:_____

* Instructions: A form/table like this may be submitted to your permitting authority to fulfill the ELGs requirement that a written report be submitted within 7 days of agreeing to participate in an INAD study. Check with your permit and permitting authority for exact reporting requirements. The first row is an example row.

Date Initiating INAD Study Participation	Name of INAD Drug Used & Dosage	Disease or Condition Intended to Treat	Method of Application
09/09/04	Oxytetracycline	For controlling columnaris in walleye	☑ Medicated feed ☐ Injection ☐ Bath treatment ☐ Other: _____
			☐ Medicated feed ☐ Injection ☐ Bath treatment ☐ Other: _____
			☐ Medicated feed ☐ Injection ☐ Bath treatment ☐ Other: _____
			☐ Medicated feed ☐ Injection ☐ Bath treatment ☐ Other: _____

* Note: This form is only an example of what a written report could look like. Facilities may use other types of existing written reports if available.

Appendix M2

Checklist for Oral Report for
INAD and Extralabel Drug Use

CHECKLIST FOR ORAL REPORT FOR INAD AND EXTRALABEL DRUG USE
(Provide an oral report to your permitting authority
within 7 days after initiating use of the drug)

* **Instructions:** This example form/table does not need to be submitted to your permitting authority. It can be used to ensure that you have fulfilled the oral reporting requirements of the ELGs. The first row is an example row.

Reported to Permitting Authority?	Name of Drug (INAD & Extralabel) Used & Reason for Use	Method of Application	First Date of Drug Use	Date Oral Report Submitted to Permitting Authority	Initials
☑	Extralabel: Erythromycin Treat bacterial infections	Injection	09/09/04	09/10/04	MJ
❑					
❑					
❑					
❑					
❑					
❑					
❑					
❑					

* Note: This checklist is only an example of a checklist that facilities could use to track oral reporting. Facilities may use existing record systems if available.

Appendix M3

Example Written Report
INAD and Extralabel Drug Use

EXAMPLE WRITTEN REPORT FOR INAD AND EXTRALABEL DRUG USE
(Submit a written report to your permitting authority
within 30 days after initiating use of the drug)

Facility Name:_____   NPDES Permit Number:_____

Name of person submitting this report:_____

Date this written report was submitted to the permitting authority:_____

* Instructions: A form like this may be submitted to your permitting authority to fulfill the ELGs requirement that a written report be submitted within 30 days after initiating use of an INAD or extralabel drug. Check with your permit and permitting authority for exact reporting requirements. The first row is an example row.

Name of Drug & Reason for Use	Date and Time of Application (start date/time end date/time)	Duration	Method of Application	Total Amount of Active Ingredient Added	Total Amount of Medicated Feed Added**
Oxytetracycline For control of columnaris in walleye	09/09/04 10:00 AM 09/13/04 10:00 AM	5 consecutive days	☑ Medicated feed ☐ Injection ☐ Bath treatment ☐ Other: _____	1 g/lb as sole ration	50 lbs
			☐ Medicated feed ☐ Injection ☐ Bath treatment ☐ Other: _____		
			☐ Medicated feed ☐ Injection ☐ Bath treatment ☐ Other: _____		
			☐ Medicated feed ☐ Injection ☐ Bath treatment ☐ Other: _____		

* This form is only an example of what a written report could look like. Facilities may use other types of existing written reports if available.
** Applies only to drugs applied through medicated feed.

Appendix M4

Checklist for Oral Report of
Failure or Damage to the Structure
of Containment Systems

CHECKLIST FOR ORAL REPORT OF FAILURE OR DAMAGE TO THE STRUCTURE OF CONTAINMENT SYSTEMS

(Provide an oral report to your permitting authority within 24 hours of the discovery of any reportable failure or damage that results in a material discharge of pollutants)

* Instructions: This example form/table does not need to be submitted to your permitting authority. It can be used as a checklist to ensure that you have fulfilled the oral reporting requirements of the ELGs. Use the following table to track failure or damage to the structure of your containment systems. The first row is an example row.

Reported to Permitting Authority?	Cause of the Failure or Damage in the Containment System	Materials Released to the Environment	Date of Release	Date Oral Report Submitted to Permitting Authority	Initials
☑	Storm/wave damage to net pen	1,000 Coho salmon	09/08/04	09/09/04	MJ
☐					
☐					
☐					
☐					
☐					
☐					
☐					
☐					
☐					

* Note: This checklist is only an example of a checklist that facilities could use to track oral reporting. Facilities may use existing record systems if available.

Appendix M5

Example Written Report
Failure or Damage to the Structure
of Containment Systems

EXAMPLE WRITTEN REPORT FOR FAILURE OR DAMAGE
TO THE STRUCTURE OF CONTAINMENT SYSTEMS
(Submit a written report to your permitting authority within
7 days of discovery of the failure or damage)

Facility Name:_____   NPDES Permit Number:_____

Name of person submitting this form:_____

Date this written report was submitted to the permitting authority:_____

* Instructions: A form like this may be submitted to your permitting authority to fulfill the ELGs requirement that a written report be submitted with 7 days of the discovery of a failure or damage to the structure of containment systems at your facility. The first row is an example row. Check with your permit and permitting authority for exact reporting requirements.

Cause of the Failure or Damage	Date Failure or Damage was Discovered	Time Elapsed Until the Failure or Damage was Repaired	Materials Released to the Environment from the Failure or Damage (Estimate)	Steps Being Taken to Prevent Reoccurrence
Broken raceway screen at quiescent zone and on standpipe	09/10/04	30 minutes	1,000 fish – Coho salmon fingerlings	1. Secure standpipe screen with pipe clamps. 2. Inspect clamps weekly for signs of corrosion or deterioration. 3. Replace clamps as necessary. 4. Inspect QZ screens weekly for signs of deterioration. 5. Replace screens as necessary.

* Note: This form is only an example of what a written report could look like. Facilities may use other types of existing written reports if available.

Appendix M6

Checklist for Oral Report of
Spills of Drugs, Pesticides, and Feed

CHECKLIST FOR ORAL REPORT OF SPILLS OF DRUGS, PESTICIDES, AND FEED

(Provide an oral report to your permitting authority
within 24 hours of any spills of drugs, pesticides, or feed)

* Instructions: This example form/table does <u>not</u> need to be submitted to your permitting authority. It can be used as a checklist to ensure that you have fulfilled the oral reporting requirements of the ELGs. Use the following table to track multiple spills throughout the year. The first row is an example row.

Reported to Permitting Authority?	Name of Material Spilled (Drugs, Pesticides, or Feed)	Quantity Spilled	Date of Spill	Date Oral Report Submitted to Permitting Authority	Initials
☑	Oxytetracycline medicated feed; broken bag spilled onto ground	50 lbs	09/09/04	09/10/04	MJ
❏					
❏					
❏					
❏					
❏					
❏					
❏					
❏					
❏					

* Note: This checklist is only an example of a checklist that facilities could use to track oral reporting. Facilities may use existing record systems if available.

Appendix M7

Example Written Report
Spills of Drugs, Pesticides, and Feed

EXAMPLE WRITTEN REPORT FOR SPILLS OF
DRUGS, PESTICIDES, AND FEED
(Submit a written report to your permitting authority within
7 days of any spills of drugs, pesticides, or feed)

Facility Name:_____   NPDES Permit Number:_____

Name of person submitting this form:_____

Date this written report was submitted to the permitting authority:_____

* Instructions: A form like this may be submitted to your permitting authority to fulfill the ELGs requirement that a written report be submitted with 7 days of a spill. The first row is an example row. Check with your permit and permitting authority for exact reporting requirements.

Name of Material Spilled (Drugs, Pesticides, or Feed)	Quantity Spilled	Where Spilled and Action Taken	Date Spilled
Oxytetracycline medicated feed	50 lbs	Bag of medicated feed broke when being moved from a pallet in the feed storage area. Contents spilled onto the floor. Swept up spilled feed and placed material in a plastic container for future use.	9/10/04

* Note: This form is only an example of what a written report could look like. Facilities may use other types of existing written reports if available.

Appendix N

Feed Conversion Ratios Log

FEED CONVERSION RATIOS LOG
FLOW-THROUGH, RECIRCULATING, AND NET PEN SYSTEMS

* **Instructions:** This example form may be used to keep track of feeding and to calculate/track feed conversion ratios. The first row is an example row. FCRs are calculated with the following equation:

$$\frac{\text{Dry weight of feed applied}}{\text{Wet weight of fish gained}}$$

Date (start date end date)	Description of Group	Total Feed Amounts (Estimate)	Weights of Animals (start weight end weight)	Weight Gained	Calculated FCR
3/20/04 10/21/04	Brooktrout stockers for Potomac River	5,275 lbs	100 lbs 4,800 lbs	4,700 lbs	1.12

Date (start date end date)	Description of Group	Total Feed Amounts (Estimate)	Weights of Animals (start weight end weight)	Weight Gained	Calculated FCR

* Note: This is only an example of what a log for tracking feeding and calculating FCRs could look like. Facilities may use existing record systems if available.

Appendix O

Spills and Leaks Log

SPILLS AND LEAKS LOG

Facility Name: _____   NPDES Permit Number: _____

* Instructions: This example form may be used to keep track of spills or leaks at your facility. You are <u>not</u> required to submit this form to your permitting authority. The first row is an example row.

Date (mm/dd/yy)	Spill or Leak	Location (as indicated on a site map)	Type of Material & Quantity	Source (if known)	Reason	Amount of Material Recovered	List of Preventative Measures Taken	Initials
09/10/04	Spill	Hatchery floor	Formalin	Storage drum	Top was not secured and the drum was knocked over	20 gallons	Spoke to all employees about the importance of securing lids to storage containers	MJ

Date (mm/dd/yy)	Spill or Leak	Location (as indicated on a site map)	Type of Material & Quantity	Source (if known)	Reason	Amount of Material Recovered	List of Preventative Measures Taken	Initials

* Note: This is only an example of what a log for tracking spills and leaks could look like. Facilities may use existing record systems if available.

March 2006

Instructions: Use this page to enter any important notes about the spills from the previous sheets.

Date of spill or leak (mm/dd/yy)	Notes
09/10/04	All employees have been instructed on the importance of securing container lids. Employees were also instructed to double-check that container lids are secured before proceeding with applications.

Appendix P1

Example Inspection and Maintenance Log
Flow-through and Recirculating Systems

STRUCTURAL MAINTENANCE
EXAMPLE INSPECTION AND MAINTENANCE LOG
FLOW-THROUGH AND RECIRCULATING SYSTEMS

Facility Name:_____   NPDES Permit Number:_____

Production or Wastewater Treatment System:_____

* Instructions: This example form may be used to keep track of routine inspections and regular maintenance of your production systems and wastewater treatment systems. Use a separate form for each production system and/or wastewater treatment system. Make sure you defined the terms "routine" and "regular" in your BMP plan. The first row is an example row.

Date Inspected	Inspector Initials	Notes (Note any problems found and maintenance performed)	Date Maintenance Performed
09/10/04	MJ	The screen at the end of raceway 2 was loose. Secured the screen to prevent it from completely coming loose.	09/12/04
09/25/04	MJ	Inspected screens in raceways. All screens were found to be in good condition.	N/A

Date Inspected	Inspector Initials	Notes (Note any problems found and maintenance performed)	Date Maintenance Performed

* Note: This is only an example of what a maintenance log could look like. Facilities may use existing record systems if available.

Appendix P2

Example Inspection and Maintenance Log
Net Pen Systems

MAINTENANCE
EXAMPLE INSPECTION AND MAINTENANCE LOG
NET PEN SYSTEMS

Facility Name:_____   NPDES Permit Number:_____

Production System:_____

* **Instructions:** This example form may be used to keep track of routine inspections and regular maintenance of your production systems. Use a separate form for each production system. Make sure you defined the terms "routine" and "regular" in your BMP plan. The first row is an example row.

Date Inspected	Inspector Initials	Notes (Note any problems found and maintenance performed)	Date Maintenance Performed
09/10/04	MJ	During a routine inspection, divers discovered a small hole in net pen 3; patched the hole.	09/10/04
09/25/04	MJ	Inspected nets – all were in good condition	N/A

Date Inspected	Inspector Initials	Notes (Note any problems found and maintenance performed)	Date Maintenance Performed

* Note: This is only an example of what a maintenance log could look like. Facilities may use existing record systems if available.

Appendix Q

Cleaning Log

CLEANING LOG
FLOW-THROUGH, RECIRCULATING, AND NET PEN SYSTEMS

Facility Name:_____   NPDES Permit Number:_____

* Instructions: This example form may be used to track cleaning of your production systems and/or wastewater treatment systems. The first row is an example row.

Date Cleaned	Cleaner Initials	Description of Component Cleaned	Notes About Cleaning
9/10/04	MJ	QZ raceways EB 1-5	Cleaned QZs and dam boards; checked and cleaned screens.

Date Cleaned	Cleaner Initials	Description of Component Cleaned	Notes About Cleaning

* Note: This is only an example of what a log for tracking cleaning of production systems and wastewater treatment systems could look like. Facilities may use existing record systems if available.

Appendix R

Record-keeping Checklist

RECORD-KEEPING CHECKLIST

* Instructions: Use the following checklist to make sure you meet all the record-keeping requirements of the CAAP ELGs. You do <u>not</u> need to submit this to your permitting authority.

❑ Records for aquatic animal rearing units documenting feed amounts and estimates of the numbers and weights of aquatic animals in order to calculate representative feed conversion ratios (can use the form in Appendix N to fulfill this requirement).

❑ Records documenting frequency of cleaning (can use the form in Appendix Q to fulfill this requirement).

❑ Records documenting frequency of inspections, maintenance, and repairs (can use forms in Appendix P to fulfill this requirement).

* Use the following checklist to see what other record-keeping forms can be used to show your permitting authority that you are meeting the reporting and BMP plan requirements of the CAAP ELGs (e.g., solids control, training). The following forms are <u>not</u> required.

❑ INAD – 7 Day Written Report (Appendix M)

❑ INAD and Extralabel – 7 Day Oral Report (Appendix M)

❑ INAD & Extralabel – 30 Day Written Report (Appendix M)

❑ Failure or Damage to the Structural Integrity of Containment Systems – 24 Hour Oral Report (Appendix M)

❑ Failure or Damage to the Structural Integrity of Containment Systems – 7 Day Written Report (Appendix M)

❑ Spills of Drugs, Pesticides, and Feed – 24 Hour Oral Report (Appendix M)

❑ Spills of Drugs, Pesticides, and Feed – 7 Day Written Report (Appendix M)

❑ Material Storage: Spills and Leaks Log (Appendix O)

❑ Cleaning Log (Appendix Q)

❑ Employee Training Log (Appendix S)

❑ Carcass Removal (Appendix T)

* Note: This checklist is only for tracking record-keeping at your CAAP facility. Facilities may use existing record systems if available.

Appendix S

Employee Training Log

EMPLOYEE TRAINING LOG

Facility Name:_____   NPDES Permit Number:_____

* Instructions: This example form may be used to track employee training at your facility.

Employee Training		Completed By: _____ Title: _____ Date: _____
Instructions: Describe the employee-training program for your facility below. The program should, at a minimum, address spill prevention and response, and proper operation and cleaning of production and wastewater treatment systems. Provide a schedule for the training program and list the employees who attend the training sessions.		

Training Topics	Brief Description of the Training Program and Materials	Schedule for Training (list dates)	Participants
Spill Prevention and Response			
Operation and Cleaning of Systems			
Feeding Procedures			
Other Topics (list):			
Other Topics (list):			

* Note: This is only an example of what an employee training log could look like. Facilities may use existing record systems if available.

Appendix T

Carcass Removal Log

CARCASS REMOVAL LOG
FLOW-THROUGH, RECIRCULATING, AND NET PEN SYSTEMS

Facility Name:_____   NPDES Permit Number:_____

* Instructions: This example form may be used to track carcass removal from your production systems and/or wastewater treatment systems. The first row is an example row.

Date	Initials	System/Group of Animals	# of Mortalities	Approx. Weight	Disposal Method	Notes
9/10/04	MJ	Brooktrout – R1–4	10	6 lbs	Composting	All from R2; appear to be from some type of infection – closely monitor remaining fish

Date	Initials	System/Group of Animals	# of Mortalities	Approx. Weight	Disposal Method	Notes

* Note: This is only an example of what a log for tracking carcass removal from your production systems and/or wastewater treatment systems could look like. Facilities may use existing record systems if available.

Appendix U

FDA Labeling

FDA: LABELS FOR NEW ANIMAL DRUGS FOR AQUACULTURE

In some cases as part of the approval process for new animal drugs, FDA may decide to include information on labels for individual aquaculture drugs in order to address the potential environmental impacts associated with the use of the drug. Drug labels may also require the user to inform the appropriate NPDES permitting authority prior to the first use of the drug. This is necessary because FDA must approve drugs for use on a nationwide basis without in-depth consideration of wastewater treatment (e.g., settling ponds) at individual facilities and local site-specific conditions such as dilution and degradation in receiving waters. The reporting requirement insures that there is appropriate oversight to determine whether effluent discharge limits are needed at individual aquaculture facilities when FDA has determined that release of a drug has the potential to cause effects on organisms in receiving waters at some locations.

Label information on aquaculture drugs may include identification of acute and chronic water quality "benchmarks" derived to address and help mitigate potential adverse effects on aquatic life resulting from drug use. These benchmarks are meant to assist NPDES permitting authorities make determinations on whether discharge limits are needed and help them set these limits, if they are needed (see below). In developing such benchmarks, FDA relies on toxicity and environmental fate information collected and generated through an environmental assessment process that is part of the overall drug approval process (Note: Environmental Assessment documents for veterinary drugs are available through the following FDA website: http://fda.gov/cvm/ea.htm). FDA's technical process for deriving water quality benchmarks is similar to that used by the U.S. EPA to develop numerical water quality criteria for the protection of aquatic life (http://www.epa.gov/waterscience/criteria/aqlife.html#guide).

Under EPA's NPDES regulations, NPDES permits must include limits necessary to achieve water quality standards under section 303 of the Clean Water Act. 40 C.F.R. § 122.44(d)(1). In cases where a State has not established a water quality criterion for a specific chemical pollutant but has established a narrative criterion (i.e, "no toxics, in toxic amounts"), the permitting authority must establish an effluent limit for a new animal drug if it is present in an effluent in concentrations that cause or has a reasonable potential to cause or contribute to an excursion above a narrative criterion. 40 C.F.R. § 122.44(d)(1)(i). In developing such limits, the permitting authority may use a calculated numeric water quality criterion derived by one of several methods, supplemented with other relevant information which may include "information about the pollutant from the Food and Drug Administration" 40 C.F.R. § 122.44(d)(1)(vi)(A). Water quality benchmarks and other information on drug labels will alert users of the potential adverse effects of drug use on aquatic life in receiving waters. This information will also provide a mechanism for alerting permit writers of the potential need to formally establish facility-specific numeric effluent limitations for aquaculture drug products as well as necessary information for complying with § 122.44(d).

Additional Information

Charles E. Eirkson III
Environmental Safety Team, HFV-103
Office of New Animal Drug Evaluation
Center for Veterinary Medicine
301-827-6653
ceirkson@cvm.fda.gov

Appendix V

SDAFS BMP Plan

BMP Plan for State Fish Hatcheries (Developed by the Southern Division of American Fisheries Society (SDAFS) Aquaculture Technical Committee

(*Italicized* text will need to be re-worded to describe each individual hatchery, and areas left blank and/or underlined will have to be filled in and may need to be updated occasionally.)

SDAFS Aquaculture Technical Committee
BMP PLAN
FOR STATE FISH HATCHERIES

Dated: _____

Facility Name: _____
Facility Address: _____

_____

_____

NPDES Number and Expiration Date: _____
Hatchery Superintendent: _____
Phone number: _____
Email address: _____

1) INTRODUCTION:

The [*fill in the name of your hatchery*] fish hatchery operates under the NPDES permit number referenced above. The NPDES permit is issued by [*fill in the US EPA or a state agency such as DENR, Division of Water Quality*). The contact person at the permit issuing authority is _____ and the contact phone number is _____ or they can be contacted by email at _____ the mailing address for the permitting authority is _____
_____. The hatchery typically produces the following types of fish, in approximately these numbers of fish and pounds of fish per year.

Species of Fish	Number of Fish	Pounds of Fish

[THIS FACILITY DESCRIPTION MAY BE TAKEN FROM ANOTHER SOURCE SUCH AS YOUR NPDES PERMIT] The hatchery consists of indoor raceways in the hatchery building and a series of re-use raceways in six parallel rows providing 46 individual outdoor culture tanks or raceways (see attached facility layout diagram). Raceways are typically 100 feet long by 8 feet wide with the first or up-stream row being twelve raceways across and each raceway being 4 feet wide.

Water is supplied through an underground pipe from a surface intake from the Wild River. Water flow is typically between 2,000 and 5,000 gallons per minute. The primary water discharge point is the outflow pipe from the lower raceway, but water can be discharged at nine locations upstream of the main water discharge point during quiescent zone cleaning. Incoming water passes through a screen at the intake and is not chemically treated or aerated before use.

2) REPORTING
The following reporting is undertaken to meet Effluent Limitations Guidelines:

1) When we **sign up** for participation in an INAD study of a reportable drug (i.e. the drug may be discharged and is not a use similar to an approved use), we submit **a written report**, which identifies the method of use, the dosage and the disease or condition being treated, **to the permitting authority within seven days**.

2) When we **use a reportable drug an oral report is given to the permitting authority within 7 days** of the use. The report includes the drug used, method of application and the reason for using the drug. **A written report is sent to the permitting authority within 30 days** of the use and the report includes the reason for treatment, date(s) and time(s) of the addition (including duration), method of application and the amount added.

3) In the event of **damage to or failure of a hatchery structure** that results in a discharge of pollutants, an **oral report is given to the permitting authority within 24 hours**. The oral report describes the cause of the failure or damage and identifies the materials released. **A written report is sent to the permitting authority within 7 days** of the problem and the written report documents the cause, the estimated time elapsed until the problem was repaired, and estimates the material released, and steps being taken to prevent a recurrence.

4) In the event of **a spill** of drugs, pesticides or feed that results in a discharge, **an oral report is given to the permitting authority within 24 hours**. The oral report describes the identity and quantity of the material spilled. **A written report is sent to the permitting authority within 7 days** of the spill and the written report describes the identity and quantity of the material spilled. Spills that are contained before they discharge to waters of the U.S. are not subject to this reporting requirement.

5) This BMP plan is finalized and being implemented at the fish hatchery. The permittee sent a letter *on [fill in date]* certifying that a BMP plan has been implemented and is available to the permitting authority upon request.

3) SOLIDS CONTROL
A) High quality feed is utilized to minimize waste. Periodically the feed formulation and manufacturing process are assessed so that the most appropriate feed is used. *Feed is either applied by hand with feed being distributed via scoop from a bucket, by belt feeder or by blower from a truck mounted automatic feeder. The feed contract specifies a high quality, extruded commercial floating trout feed with a minimum protein content of 42% and minimum 16% fat content for all trout grower and finisher feeds.* Feeding is adjusted to meet requirements of the fish based

on the number of fish, size of the fish, feeding response of the fish, and the temperature of the water. Feed consumption is visually monitored and when the floating feed is not readily consumed by the fish the feed rate is adjusted to prevent overfeeding either during that feeding or for the next feeding that is to occur.

B) Fish inventories are continuously updated based on stocking rates, records of mortalities and fish growth. The following measures are taken to minimize solids discharge during grading, harvesting and inventorying of fish: 1) fish are not fed for 24 hours before handling, and 2) screens and quiescent zones are cleaned before fish are handled (etc). Raceways are stocked with proper numbers of fish to facilitate movement of solids through the raceway system. *Feed rates are adjusted weekly based on fish inventory and other considerations mentioned above. Fish inventory is reported monthly. We use feeding records and inventory records to calculate feed conversion on a monthly and annual basis. Feed conversion ratios are reported in monthly and annual reports. A physical inventory of the fish based on weight of fish in each raceway and size (from sub-sampling) of the fish is conducted as needed, but at least once each year.*

C) There is a perimeter fence around the raceways to keep wildlife from capturing and removing fish. Quiescent zones are maintained in the downstream end of each raceway by screening the fish out of the area. Quiescent zones are four feet long and are cleaned at least once weekly on a rotating basis. Solids from each quiescent zone are brushed out and flushed through the discharge system to the stream through the drain at the bottom of each quiescent zone. No more than one section of quiescent zones (one fourth of the facility) is cleaned each day.

D) Two settling ponds collect solids during cleaning of the quiescent zones. Solids are removed from the settling ponds every other month by a septic tank pump truck, and land applied at agricultural rates.

E) Trout mortalities are collected from each raceway at least twice per week. The carcasses are disposed of on site well away from receiving waters so that there is no chance of mortalities making their way to receiving waters. Mortalities are collected before they deteriorate and discharge back to the river.

A disposal log is maintained at the hatchery and updated each time solids (typically dead fish) are removed. The log contains:
1. *date of disposal*
2. *area where solids were applied*
3. *amount of solids applied*
4. *initials of applicator*

4) MATERIALS STORAGE
A) Employees are trained in proper handling and storage of materials used in the hatchery. The facility maintains a list of all materials that require special handling in the hatchery together with relevant MSDS sheets. A spill response plan is attached as Appendix A. Particular materials of concern are:
- *Feed in bags*

- *Bulk feed*
- *Medicated feed*
- *Therapeutants (formalin, salt, anesthetics, etc.)*
- *Fuels and lubricants*
- *Disinfectants*

B) New employees and existing staff are trained to avoid any spills that could enter public waters, and properly dispose of spilled substances. The hatchery superintendent schedules training on an annual basis to update hatchery staff. The training addresses each of the types of materials of concern listed above (Appendix A, Spill Response Plan).

5) STRUCTURAL MAINTENANCE

A) New employees and existing staff are trained to be alert to leakage from or deterioration of production and waste storage facilities when feeding or working around the raceways. When any malfunctioning of facilities is observed it is immediately reported to the superintendent who takes appropriate action to correct the situation. In addition, as part of the annual training, an annual inspection of production and waste storage facilities is conducted.

B) Maintenance of intake screens, raceway screens, *LHOs and other facility systems* is done on a daily basis (as described by the manufacturer's specifications if available) during feeding and other activities. Feed storage bins and areas are kept clean and pest free on a daily basis. *A notebook that lists maintenance on vehicles and equipment is maintained on site.*

6) RECORD KEEPING

A) Feeding records are maintained daily and feed usage is summarized and reported monthly. *Monthly reports are stored for a period of at least five years. Monthly reports are compiled into an annual report, which summarizes production and feeding data including FCR, total feed usage, and total production in numbers and pounds of fish.* All records are available on site upon request.

B) Forms that track cleaning, inspections, maintenance, waste disposal, training and repairs are kept on site, compiled on an annual basis and kept with the annual reports.

7) TRAINING

A) Once each year, *during January*, the hatchery superintendent arranges a *half-day* training session for all employees at the hatchery. During the training session, the BMP plan is reviewed in detail and each section is discussed. *Other operational plans such as the fire exit plan, safety plan, spill response plan, stocking procedures and hatchery operational procedures are also reviewed.*

B) Each new employee at the hatchery is given an orientation that includes a detailed review of and training in the BMP plan. This training is conducted within the first two weeks that the employee is on the job.

8) FACILITY LAYOUT DIAGRAM

A layout diagram of the hatchery facility is attached indicating where water intakes are located, where water discharges are located, where feed and chemical storage is located, where culture facilities are located, and where waste storage facilities (including disposal for fish carcasses) are located.

APPENDIX A -- SPILL RESPONSE PLAN

1) IMMEDIATE RESPONSE

- Don't panic. **Call 911 if public safety is threatened.** Get help on site and call for more help if necessary.
- Define the problem (leaking valve, broken container, overflow, etc.).
- Assess risks (where will spill go and will it enter your water discharge).
- Keep people safe. Away from the spill, upwind or evacuate as necessary.
- Stop the source of spill if possible, safe and necessary.
- Stop sources of ignition if relevant (shut off motors, engines, no-smoking, etc.).
- Contain the spill if safe and possible.
  - o Collect the spill in a bucket or drip pan.
  - o Block the spill from spreading or getting into the water (build dike or block with sandbags, etc.).
- Call for help.

2) STABILIZATION

- Clean up the spill safely if you can or arrange for a contractor to clean it up.
- Log the spill, and review Spill Response plans and update as needed.

3) IMPORTANT PHONE NUMBERS AND CONTACT INFORMATION

- 911
- Agency Contacts
  - o Hatchery Superintendent – Home_____ Work_____
  - o Production Coordinator - Home_____ Work _____
  - o Regional Supervisor - Home _____ Work _____
  - o Program Coordinator - Home _____ Work _____
- NPDES Permit Liaison: Name _____ Phone Number _____
- Other Government Agencies, Local Officials, and Neighboring Facilities.

4) PREVENTION

Preventative measures and procedures are listed below.

- *Chemical storage room or cabinet with lip to prevent spills in storage room.*
- *Containment barrier around fuel tanks, and overfill protection on generator tank.*
- *Material Safety Data Sheets are maintained for all chemicals used.*
- *Security fence, locks, and lights.*
- *Inspection logs and procedures.*
- *Labeling of tanks and containers.*
- *Diagram of site, storage areas, and exit plans.*

5) PREPAREDNESS

- Available equipment and supplies that can be used to control spills:
  - o *Shovels and brooms*
  - o *Empty buckets and drums*
  - o *Plastic sheeting and plastic bags*
  - o *Sand bags and absorbent materials*
- *Spill containment materials are located in the feed storage room.*
- *Annual training and new employee training includes spill response training.*

www.ingramcontent.com/pod-product-compliance
Lightning Source LLC
Chambersburg PA
CBHW080634180526
45168CB00008B/3160